P9-ASA-585

Other titles of interest

Seven Years of MANIFOLD: 1968–1980
Ian Stewart & John Jaworski (Editors)

The Collected Letters of Colin MacLaurin
Stella Mills (Editor)

In the Shiva Mathematics Series

Categories, Bundles and Spacetime Topology
C.T.J. Dodson

Group Tables
A.D. Thomas & G.V. Wood

Manifolds of Differentiable Mappings
P.W. Michor

University of Strathclyde Seminars in Applied Mathematical Analysis: Vibration Theory
G.F. Roach (Editor)

Notes on Real and Complex C*-Algebras
K.R. Goodearl

Applications of Combinatorics
R.J. Wilson (Editor)

Applied mathematical analysis: vibration theory

University of Strathclyde Seminars in

Applied mathematical analysis: vibration theory

G F Roach (Editor)
University of Strathclyde

Shiva Publishing Limited

SHIVA PUBLISHING LIMITED
4 Church Lane, Nantwich, Cheshire CW5 5RQ, England

Hardback edition available in North America from:
BIRKHAUSER BOSTON, INC.
P.O. Box 2007, Cambridge, MA 02139, USA

British Library Cataloguing in Publication Data

Roach, G.F.
Applied mathematical analysis: vibration theory.
— (Shiva mathematics series; 4)
1. Vibration — Congresses
I. Title
531'.32 QA865

ISBN 0-906812-12-7
ISBN 0-906812-11-9 Pbk

Printed in Great Britain by Devon Print Group, Exeter

Contents

Watson transform and high frequency scattering

B.D. SLEEMAN

§1

It is well accepted that certain physical theories are short-wave limits of more general theories. Thus for example ray optics is the short wave limit of wave optics. It is natural to expect to be able to deduce ray optics by a rigorous mathematical limiting process from the theory of wave optics. However it has never been proved in general that the wave solution of the general exterior boundary value problem for Helmholtz equation tends to the ray limit, still less that the assumed form of the complete asymptotic expansion is correct; i.e. that the assumed expansion is an asymptotic approximation to the exact solution as the wave length tends to zero.

Although these results have not been proved in the general case considerable progress has been made for the so called "canonical problems" wherein the boundary of the geometrical configuration is a level surface in a coordinate system for which Helmholtz equation separates. Separation of variables methods can then be used to obtain an explicit solution in the form of a single series (in two dimensions) or a double series (in three dimensions). Unfortunately the resulting series converge very slowly at short wavelengths and are therefore virtually useless. However the Watson transformation (12) can often be applied, for two dimensional problems, to transform the single series into a rapidly convergent series. Having done this the resulting solution may be used to go a long way in laying a rigorous foundation to two dimensional short wave asymptotics. Certain aspects of this will be discussed in later sections of this paper.

In order to make progress in the case of three dimensional problems it is desirable to generalise the Watson transformation to double series. This is the central feature of this paper.

In Section 2 we review the Watson transformation and discuss a number of its more important implications in Section 3. Section 4 motivates in an informal way the generalisation of the Watson transformation to three dimensional problems and briefly considers the problem of scattering by a prolate spheroid. It includes the residues series which may be obtained with the use of the generalised Watson transformation. A fairly detailed application to the scattering problem for a sphere is considered in Section 5 which, it is conjectured, gives an insight into future developments.

§2 The Watson Transformation

Suppose sound waves are emitted towards infinity by a time periodic source situated at $\theta = 0$ on the circle $r = a$, where the period is $2\pi/\omega$ and the speed of sound is c. If we denote the wave length ω/c by k then in the assumed polar coordinate system (r,θ) the velocity of potential $\phi(r,\theta)e^{-i\omega t}$ satisfies the equation

$$\left(\frac{\partial^2}{\partial r^2} + \frac{1}{r}\frac{\partial}{\partial r} + \frac{1}{r^2}\frac{\partial^2}{\partial \theta^2} + k^2\right)\phi = 0, \tag{2.1}$$

subject to the boundary condition

$$\frac{\partial \phi}{\partial r} = \frac{2i}{a}\delta(\theta) \qquad \text{on } r = a \tag{2.2}$$

and the radiation condition

$$r^{\frac{1}{2}}\left(\frac{\partial}{\partial r} - ik\right)\phi \to 0 \quad \text{as } r \to \infty . \tag{2.3}$$

It is a standard exercise in the method of separation of variables to show that the velocity potential $\phi(r,\theta)$ can be written as the single infinite series

$$\phi(r,\theta) = \frac{i}{\pi ka}\left[\frac{H_o^{(1)}(kr)}{H_o^{'(1)}(ka)} + 2\sum_{m=1}^{\infty}\frac{H_m^{(1)}(kr)}{H_m^{'(1)}(ka)}\cos m\theta\right] \tag{2.4}$$

where $H_m^{(1)}(z)$ denotes the Hankel function of the first kind and ' denotes differentiation with respect to the argument.

When $N = ka$ is small (2.4) provides an adequate series solution to the problem. However when N is large the series (2.4) converges so slowly as to be virtually useless. It is precisely this situation in which the Watson transformation is used. To be precise we consider the series (2.4) as the resulting residue series obtained from the evaluation of an appropriate contour integral in the complex plane. The desired integral takes the form

$$\phi(r,\theta) = \frac{1}{\pi ka}\int_C \frac{H_\nu^{(1)}(kr)}{H_\nu^{(1)\prime}(ka)} \frac{\cos \nu(\pi - \theta)}{\sin \nu\pi} d\nu \tag{2.5}$$

where C is the contour shown in figure 1 where the radius of the arc BC is such that it never passes through a zero of $\sin \nu\pi$.

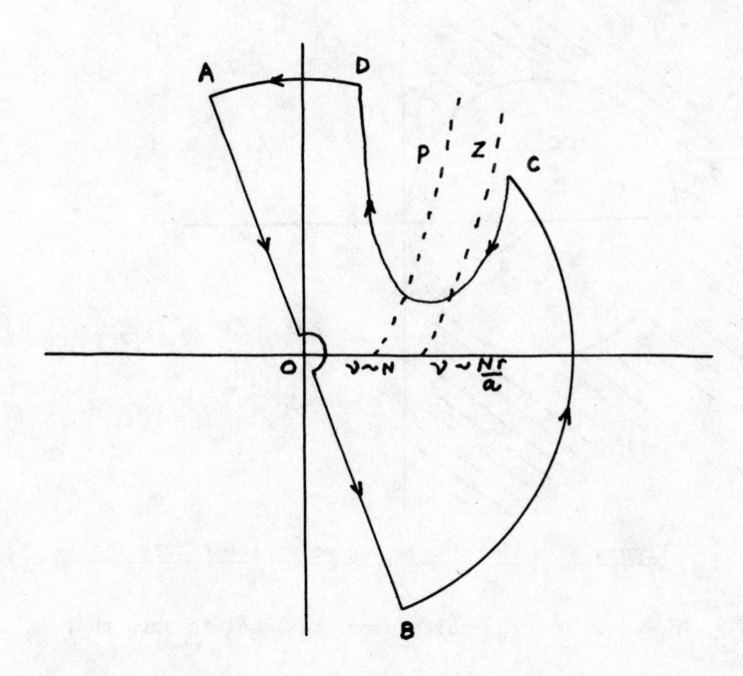

Figure 1. The complex ν-plane. $(N = ka)$. The zeros of $H_\nu^{(1)(\prime)}(ka)$ lie near the pole locus P, the zeros of $H_\nu(kr)$ lie near the zero locus Z. Near $\nu = N$ the curve DC passes between the first zero of $H_\nu^{\prime(1)}(ka)$ and the real ν-axis.

Since the integrand in (2.5) is an odd function of ν there is no contribution to the integral along AB. Also for $r > a$, $0 < \theta < 2\pi$ it can be shown that the contributions from the circular arcs BC and DA vanish in the limit as $R \to \infty$. The poles inside the contour are at the zeros of $\sin \nu\pi = 0$; that is at $\nu = 1,2,3,\ldots$. (There is also a contibution from the indentation $\nu = 0$).

From the theory of residues it follows that the representation (2.5) and the series (2.4) are equivalent.

Now the contour DC may also be regarded as a closed contour enclosing the zeros $H_\nu^{(1)'}(ka)$ provided the integrand in (2.5) tends to zero suitably on a sequence of contours not passing through the zeros of $H_\nu^{(1)'}(ka)$. It can be shown Ursell (9), that it does so in a region including the shadow region $r \cos \theta < a < r$ of the circle depicted in figure 2.

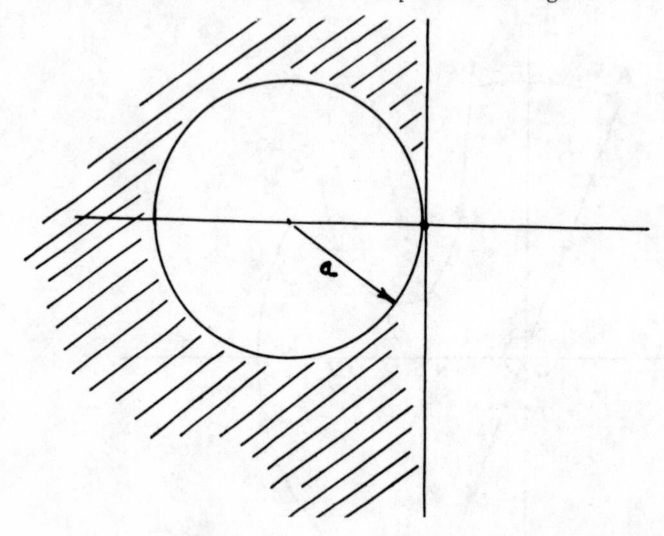

Figure 2. The shadow region of the circle $r = a$

Once again we can use the calculus of residues to show that

$$\phi(r,\theta) = -\frac{2i}{ka} \sum_{\nu=\nu(m,ka)} \frac{H_\nu^{(1)}(kr)}{\frac{\partial}{\partial\nu} H_\nu^{(1)'}(ka)} \frac{\cos \nu(\pi - \theta)}{\sin \nu\pi}, \qquad (2.6)$$

where the summation is taken over the values of $\nu(m,ka)$ of ν that satisfy $H_\nu^{(1)'}(ka) = 0$ ranked in order of increasing imaginary parts.

From the asymptotic behaviour of the derivative of the Hankel function $H_\nu^{(1)}(ka)$ it can be shown that the values of $\nu(m,ka)$ are given by

$$\nu(m,ka) = ka + 2^{-1/3} e^{\pi i/3} (ka)^{1/3} |a'_m| + \ldots ,$$

where a'_m is the m-zero of the derivative of the Airy function Ai(z).

Now from the uniform asymptotic behaviour of the Hankel functions involved it is seen that the representation (2.6) is rapidly convergent in the shadow of the circle $r = a$. In fact just the first few terms are usually sufficient. Full details leading to the above results are to be found in Ursell (9,11).

A similar analysis may be performed in the case of short wave asymptotics for the ellipse (see Leppington (3,4)) and in certain radially symmetric problems for the prolate spheroid (see Sleeman 7). In these situations the analysis is not without surprises and is further complicated by the need to invoke the uniform asymptotic behaviour of Mathieu functions or spheroidal wave functions.

For each of these essentially two-dimensional problems the results are in precise agreement with those predicted by the geometrical theory of diffraction developed in the 1950's and 60's by J.B. Keller and his school at the Courant Institute. (See R.M. Lewis, N. Bleistein and D. Ludwig 5.)

Suppose we have a closed curve S which is not a level surface in which Helmholtz equation is separable, clearly separation of variables methods can no longer be applied and consequently no analogue of the Watson transformation is available to study short-wave asymptotics. Nevertheless there is a wide class of important and fundamental problems which can be resolved in the two-dimensional case and, as we shall see in §3, in which the Watson transformation plays a significant role.

§3 Validity of the geometrical theory of diffraction and the rigorous foundation of short-wave asymptotics

The first attack on establishing the validity of the geometrical theory of diffraction for non-canonical problems is that of Bloom and Matkowsky (1). In that work they considered scattering of a wave from an infinite line source by an infinitely long cylinder C say. The line source is parallel to the axis of C and the cross section Ω of the cylinder is smooth, closed and convex. Furthermore Ω is formed by joining a pair of smooth convex arcs to a circle Ω_0, one on the illuminated side and one on the dark side so that Ω is circular near the points of diffraction. By rigorous argument Bloom and Matkowsky derived the asymptotic behaviour of the field at short wave lengths in the deep shadow of Ω. The leading term of the asymptotic expansion is shown to be the field predicted by the geometrical theory of diffraction.

A genuine extension of these results was obtained by Sleeman (8). The extension is based on the fundamental observation that, given an arbitrary closed convex curve, for any two points on the curve, we can construct a one-parameter family of ellipses passing through the two points and having two-point contact with the given curve at these points. This observation suggests that we consider Ω to be formed by joining a pair of smooth convex arcs to an ellipse Ω_0, one on the illuminated side and one on the dark side, so that Ω is elliptic near the points of diffraction.

The analysis of this problem falls naturally into two parts. Firstly we consider the scattering of a circular wave by a smooth convex curve Ω_1 which is elliptic on its dark side and also near the points of "diffraction". Thus Ω_1 may be considered as being formed by pasting a convex curve B_1 to the illuminated side of an ellipse Ω_0. (See Figure 3.) It is established that if $\underset{\sim}{r}$ is a point in the deep shadow of Ω_1 and k is large then the total field $U_1(\underset{\sim}{r};k)$ is given asymptotically by $U_0(\underset{\sim}{r};k)$ were $U_0(\underset{\sim}{r};k)$ represents the total field at $\underset{\sim}{r}$ due to the scattering of a circular wave by Ω_0. The uniform asymptotic expansion obtained from $U_1(\underset{\sim}{r};k)$ is in a form predicted by Lewis, Bleistein and Ludwig (5). From this the non-uniform series expansions of the extended geometrical

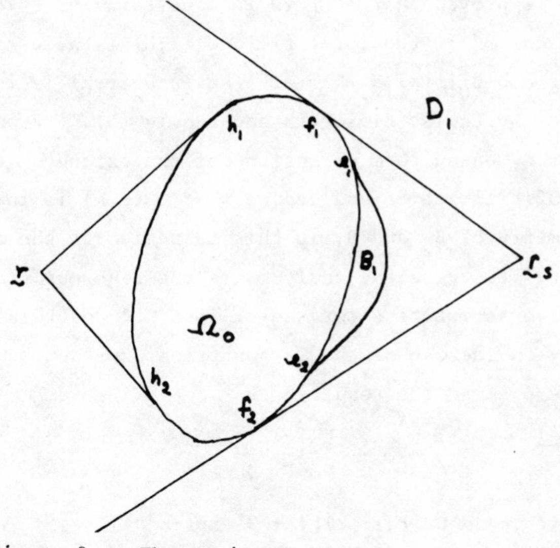

Figure 3. The region $\Omega_1 \cup D_1$

theory of diffraction (13) is obtained. The insensitivity of $U_1(\underset{\sim}{r};k)$ to the geometry of B_1 as $k \to \infty$ is that predicted by the geometrical theory of diffraction.

In the more general case where a circular wave is scattered by a smooth convex curve Ω_2 which coincides with Ω_0 only near the points of diffraction, we imagine Ω_2 as being formed by pasting a convex hump B_2 to the dark side of Ω_1 (see Figure 4).

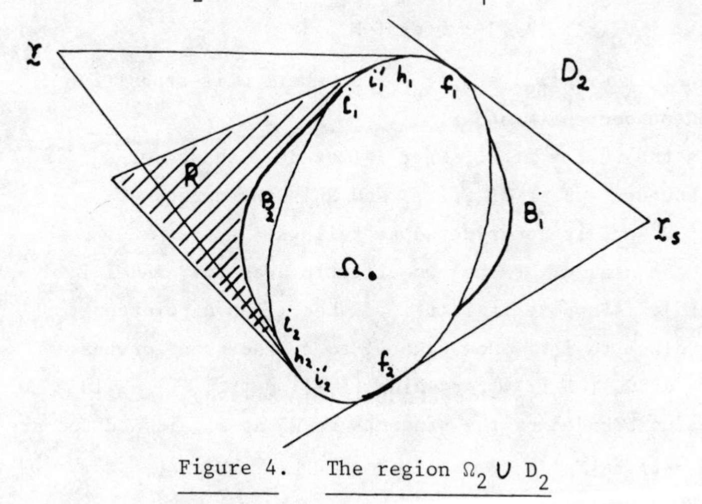

Figure 4. The region $\Omega_2 \cup D_2$

Here it is proved that if $\underset{\sim}{r}$ is in a certain subregion of the deep shadow of Ω_2 the total field $U_2(\underset{\sim}{r};k)$ is also asymptotic to the field $U_0(\underset{\sim}{r};k)$ as $k \to \infty$. Again $U_2(\underset{\sim}{r};k)$ is in the form predicted by Lewis, Bleistein and Ludwig (5). From this we derive the non-uniform expansions of the extended geometrical theory of diffraction and deduce that $U_2(\underset{\sim}{r};k)$ is insensitive to the geometry of B_2 and B_1 in this subregion of the deep shadow.

Central to the above analysis is the rigorous treatment of the short-wave asymptotics for the ellipse Ω_0, and this depends entirely on the use of the Watson transformation.

To be precise the results are

Theorem 1

As $k \to \infty$

$$U_1(\underset{\sim}{r},\underset{\sim}{r}_s,k) = U_0(\underset{\sim}{r},\underset{\sim}{r}_s,k)[1 + O(\exp(-k^{\frac{1}{2}}\sigma))],$$

uniformly in $\underset{\sim}{r}$, $\underset{\sim}{r} \in S_1^<(\underset{\sim}{r}_s)$, where σ is a positive constant independent of $\underset{\sim}{r}$ and k, $U_0(\underset{\sim}{r};\underset{\sim}{r}_s;k)$ is the solution to the scattering problem for the ellipse Ω_0 and $S_1^<(\underset{\sim}{r}_s)$ is the "deep" shadow of Ω_1 defined as follows. $\underset{\sim}{r} \in S_1(\underset{\sim}{r}_s)$, the shadow of Ω_1, if and only if $\underset{\sim}{r} \in D_1 \cup \Omega_1$ and the straight line through $\underset{\sim}{r}$ and the source point $\underset{\sim}{r}_s$ cuts Ω_1 at two distinct points. $S_1^<(\underset{\sim}{r}_s)$ is any closed bounded subset of $S_1(\underset{\sim}{r}_s)$.

Theorem 2

As $k \to \infty$

$$U_2(\underset{\sim}{r};\underset{\sim}{r}_s;k) = U_0(\underset{\sim}{r};\underset{\sim}{r}_s;k)[1 + O(\exp(-k^{\frac{1}{2}}\mu)]$$

uniformly in $\underset{\sim}{r}$, $\underset{\sim}{r} \in S_2^-(r_s) = S_2^<(r_s) - R$, where μ is a positive constant independent of k and $\underset{\sim}{r}$.

$S_2(\underset{\sim}{r}_s)$ is the shadow of Ω_2, $S_2^<(\underset{\sim}{r}_s)$ (the deep shadow of Ω_2) is any closed bounded subset of $S_2(r_s)$ and R is the region of influence of B_2 and is constructed as follows.

Consider the smaller of the two elliptic arcs f_1i_1 and f_2i_2 of $\Omega_2 \cap S_2(r_s)$. Suppose it is f_1i_1. Let i_1' be a point on f_1i_1 arbitrarily close to i_1. Now take i_2' to be the point on the elliptic arc of $\Omega_2 \cap S_2(\underset{\sim}{r}_s)$ for which $|f_1i_1'| = |f_2i_2'|$. R is then the region bounded by the tangents to Ω_2 at i_1', i_2' and the arc $i_1'i_1i_2i_2'$ of $\Omega_2 \cap S_2(\underset{\sim}{r}_s)$.

For proofs of these results and other developments we refer to (8).

In 1966 Ursell (10) discussed a number of important problems arising in the rigorous foundations of short wave asymptotics. As a typical example consider the two-dimensional exterior Neumann problem of acoustics in which the normal velocity $V(s;k)$ is prescribed on a smooth convex curve. The boundary values of the velocity potential $\phi(s;k)$ satisfy the integral equation

$$\phi(s';k) - \frac{1}{2}i\int_{\partial D} \phi(s;k)\frac{\partial}{\partial n}G(s,s';k)\,ds = -\frac{1}{2}i\int_{\partial D} V(s,k)G(s,s';k)\,ds$$

where G is an arbitrary fundamental solution of the Helmholtz equation and having a source singularity at s'. Since there are infinitely many fundamental solutions there are infinitely many integral equations with the same unique solution $\phi(s;k)$. Ursell bases a rigorous argument on which the asymptotic behaviour of $\phi(s';k)$ $(k\to\infty)$ can be found on choosing G in such a way that $\frac{\partial}{\partial n}G(s,s';k) \to 0$ as $k \to \infty$. The construction given makes use of canonical solutions for Helmholtz equation and these in turn are treated via the Watson transformation.

In this way Ursell opens up the possibility of

(a) justifying ray theory to an arbitrarily high degree of approximation in the illuminated region outside a smooth convex curve;

(b) treating the associated Dirichlet problem;

(c) applying similar techniques to two-dimensional surface wave problems.

In all these problems the Watson transformation plays a crucial role.

It is natural to ask similar questions to those mentioned here but in a three-dimensional setting. In order to do this it is not unreasonable to expect the analogous theories to depend heavily on a generalisation of the Watson transformation to these higher dimensional problems. The canonical problems likely to require investigation here include diffraction by prolate spheroids, ellipsoids and possibly paraboloids of revolution. The manner in which such a generalisation is to be achieved is considered in the following section.

§4 The Generalised Watson Transformation

In order to derive the appropriate generalisation of the Watson transformation we must go back to the problem of Section 2 and formalise developments leading to the contour integral representation (2.5). This is done on the observation that the factor $-\dfrac{\cos\nu(\pi-\theta)}{\sin\nu\pi}$ is the Green's function for the problem

$$\frac{d^2y}{d\theta^2}+\nu^2y=-\delta(\theta)$$

y - 2π periodic.

Similarly the factor $-\dfrac{2i}{ka}\dfrac{H_\nu^{(1)}(kr)}{H_\nu^{'(1)}(ka)}$ is the Green's function for the problem

$$r^2\frac{d^2y}{dr^2}+r\frac{dy}{dr}+(r^2k^2-\nu^2)y=-\delta(r-a)$$

$$\frac{dy}{dr}=0,\quad r=a,$$

and y - satisfies the Sommerfeld radiation condition. If we denote

$$G_\theta(\theta,0,\nu)=-\frac{\cos\nu(\pi-\theta)}{\sin\nu\pi},$$

and

$$G_r(r,a,k,\nu)=-\frac{2i}{ka}\frac{H_\nu^{(1)}(kr)}{H_\nu^{'(1)}(ka)}$$

then an analysis paralleling that of Sleeman (6) we find that

$$\phi(r,\theta)=\frac{1}{2\pi i}\int_C G_r(r,a,k\ ,\nu)G_\theta(\theta,0,\nu)d\nu, \tag{4.1}$$

where C is precisely the contour depicted in Figure 1.

Suppose we now wish to discuss short-wave asymptotics for a prolate spheroid illuminated by a point source. In spheroidal coordinates, related to Cartesian coordinates by

$$x=c(\xi^2-1)^{\frac12}(1-\eta^2)^{\frac12}\cos\phi,$$

$$y=c(\xi^2-1)^{\frac12}(1-\eta^2)^{\frac12}\sin\phi,$$

$$z=c\xi\eta,$$

$$\xi\in(1,\infty),\ \eta\in(-1,1),\ \phi\in(0,2\pi),$$

we consider the spheroid to be represented by the level surface $\xi = \xi_0$ and the point source to be prescribed at $(\bar{\xi},\bar{\eta},\bar{\phi})$.

The procedure now is to construct the following set of Green's functions

(a) $G_1(\phi,\bar{\phi},\mu)$ satisfying

$$\frac{d^2G_1}{d\phi^2} + \mu^2 G_1 = -\delta(\phi - \bar{\phi}), \quad 0 \leq \phi,\ \bar{\phi} \leq 2\pi \tag{4.3}$$

G_1 - 2π periodic.

(b) $G_2(\eta,\bar{\eta};\mu,\lambda)$ satisfying

$$\frac{d}{d\eta}[(1-\eta^2)\frac{dG_2}{d\eta}] - \{-k^2c^2(1-\eta^2) + \mu^2(1-\eta^2)^{-1} + \lambda\}G_2 = -\delta(\eta - \bar{\eta}).$$

G_2 - regular at $\eta = \pm 1$ and single valued except possibly at the point $\eta = \bar{\eta}$.

(c) $G_3(\xi,\bar{\xi};\mu,\lambda)$ satisfying

$$\frac{d}{d\xi}[(\xi^2 - 1)\frac{dG_3}{d\xi}] - \{-k^2c^2(\xi^2 - 1) + \mu^2(\xi^2 - 1)^{-1} - \lambda]G_2 = -\delta(\xi - \bar{\xi})$$

$$\frac{dG_3}{d\xi} = 0 \quad \text{on } \xi = \xi_0 > 1 \tag{4.4}$$

and G_3 - satisfies the Sommerfeld radiation condition.

Notice that G_1 is analogous to the Green's function $G_\theta(\theta,0;\nu)$ constructed above. The function G_2 involves spheroidal wave functions whose **behaviour** is related to that of the associated Legendre function $P^{\mu}_{\nu}(\eta)$, $(\lambda = \nu(\nu + 1))$ and the function G_3 involves spheroidal wave functions whose behaviour is related to that of the spherical Hankel functions.

Following the development in (6) it can be shown that the velocity potential

$$\Phi(\xi,\eta,\phi;\bar{\xi},\bar{\eta},\bar{\phi}) = \frac{1}{4\pi^2}\int_{\Gamma_\mu}\int_{\Gamma_\lambda} G_1(\phi,\bar{\phi},\mu)G_2(\eta,\bar{\eta},\lambda,\mu)\,G_3(\xi,\bar{\xi},\lambda,\mu)\,d\lambda d\mu. \tag{4.5}$$

where $\Gamma_\mu,\Gamma_\lambda$ are contours in the complex μ and λ planes respectively. The key to the whole matter now is to evaluate (4.5) as a double residue series particularly suited to short

at short wave lengths. From (6) it turns out that there are several choices that can be adopted for the contours Γ_μ and Γ_λ. For example if Γ_μ is a contour in the complex μ plane of the same form as that in Figure 1 then it encloses the poles of G_1 which occur at the zeros of sin $\mu\pi$. That is the points $\mu = 1,2,\ldots$. Now if for each fixed $\mu = n$, Γ_λ is chosen to be a closed contour enclosing the poles of G_2 (they are simple poles) then (4.5) can be evaluated as a residue series which is precisely the double series expansion of Φ obtained by the usual separation of variables. In other words it is analogous to (2.4) and consequently useless at short wave lengths.

Another choice for $\Gamma_\mu,\Gamma_\lambda$ is as follows:- Take Γ_μ to be the contour shown in Figure 1 and then for each integer n choose Γ_λ to be the contour surrounding the simple poles of G_3. These poles behave for large k in a manner similar to the zeros of the Hankel function $H_\nu^{'(1)}$(ka). In fact Γ_λ will have the appearance of a succession of curves similar to the arc DC in figure 1 each indexed by n and each excluding the poles arising from G_2.

It is conjectured that on evaluating the resulting residue series in this case the double series obtained is rapidly convergent in the shadow of the spheroid (see Section 5).

Finally it may be possible to consider the choice $\Gamma_\mu \Gamma_\lambda$ to surround in some sense the singularities of G_2 and G_3. However it is not known at present whether such a choice is valid or whether the resulting residue series is a solution to the scattering problem or even has any useful convergence properties at short wave lengths.

§5 Short wave diffraction by a sphere

In order to support the conjecture of §4 let us consider the problem of scattering of an arbitrary point source by a sphere of radius a. In spherical polar coordinates (r,θ,ϕ) the sphere is characterised by r = a and the source is taken without loss of generality to be situated at $(a,\bar{\theta},0)$.

If we follow an analysis similar to that leading to (4.5) we find that we have to consider integrals of the form

$$I = \int_{\Gamma_\mu}\int_{\Gamma_\nu} \frac{\Gamma(\nu + \mu - 1)}{\Gamma(\nu - \mu + 1)} \frac{P_\nu^{-\mu}(\cos \theta_>)P_\nu^{-\mu}(-\cos \theta_<)}{\sin(\nu - \mu)\pi} \frac{\cos \mu(\pi - \phi)}{\sin \mu\pi}$$

$$\times \frac{h_\nu^{(1)}(kr)}{h_\nu^{'(1)}(ka)} d\mu d\nu \tag{5.1}$$

where $r \geq a$, $\phi \in (0,2\pi)$, $\theta \in (0,\pi)$, $\theta_> = \max(\theta,\bar{\theta})$, $\theta_< = \min(\theta,\bar{\theta})$ and $h_\nu^{(1)}$ denotes the spherical Hankel function of the first kind and of order ν. Also $P_\nu^\mu(z)$ denotes the associated Legendre function.

If we choose Γ_μ to be the contour of figure 1 and Γ_ν to be a closed contour surrounding the zeros of $\sin(\nu - \mu)\pi$ for each integer value of μ; then we would obtain the usual separation of variables solution to our problem.

Suppose we proceed to integrate (5.1) with respect to μ along the contour Γ_μ then

$$I = \int_{\Gamma_\nu} \frac{P_\nu(\cos \theta_<)P(- \cos \theta_>)}{\sin \nu\pi} \frac{h_\nu^{(1)}(kr)}{h_\nu^{'(1)}(ka)} d\nu$$

$$+ 2 \sum_{m=1}^{\infty} \int_{\Gamma_\nu} \frac{\Gamma(\nu + m + 1)}{\Gamma(\nu - m + 1)} \frac{P_\nu^{-m}(\cos \theta_<)P_\nu^{-m}(-\cos \theta_>)}{\sin(\nu - m)\pi} \cos m(\pi - \phi) ,$$

$$\times \frac{h_\nu^{(1)}(kr)}{h_\nu^{'(1)}(ka)} d\nu$$

which on using the addition formula for the Legendre function (2) we have

$$I = \int_{\Gamma_\nu} \frac{h_\nu^{(1)}(kr)}{h_\nu^{'(1)}(ka)} P_\nu(\cos\theta_< \cos\theta_> + \sin\theta_< \sin\theta_> \cos(\pi - \phi))d\nu. \tag{5.2}$$

One can now proceed as we have outlined in section 2 of this paper to show that when (5.2) is evaluated as a residue series arising from the zeros of $h_\nu^{'(1)}(ka)$ we obtain an expansion which is rapidly convergent at short wave lengths in the shadow of the sphere. Thus in this simple case the conjecture of section 4 is confirmed when we observe that in the notation of that section G_1 is a multiple of the factor $\frac{\cos \mu(\pi - \phi)}{\sin \mu\pi}$, G_2 is a multiple of the factor $\frac{\Gamma(\nu + \mu + 1)}{\Gamma(\nu - \mu + 1)} \frac{P_\nu^{-\mu}(\cos \theta_>)P_\nu^{-\mu}(-\cos \theta_<)}{\sin(\nu - \mu)\pi}$ and G_3 is a multiple of the factor $\frac{h_\nu^{(1)}(kr)}{h_\nu^{'(1)}(ka)}$.

Finally it would be of interest to examine the possibility in this simple situation, of the validity or otherwise of choosing Γ_μ Γ_ν to be closed contours surrounding the poles of G_2 and G_3.

Clearly there is considerable food for thought in the problems of this paper for the analyst and the applied mathematician and their successful resolution would lead to a considerable advance in our understanding of short wave phenomena.

References

[1] C.A. Bloom and B.J. Matkowsky. On the validity of the geometrical theory of diffraction by convex cylinder. Arch. Rat. Mechs. Anal. **33** (1969) 71-90.

[2] A. Erdelyi, et al. Higher transendental functions, Vol. 2, McGraw-Hill, New York, 1953.

[3] F.G. Leppington. Creeping waves in the shadow of an elliptic cylinder. J. Inst. Maths. Applics. **3** (1967), 388-402.

[4] F.G. Leppington. On the short-wave asymptotic solution of a problem in acoustic radiation. Proc. Camb. Philos. Soc. **64** (1968) 1131-1150.

[5] R.M. Lewis, N. Bleistein and D. Ludwig. Uniform asymptotic theory of creeping waves. Comm. Pure Appl. Math. **20** (1967) 295-327.

[6] B.D. Sleeman. Integral representations associated with high-frequency scattering by prolate spheroids. Quart. J. Mech. Appl. Math. **22** (1968) 405-426.

[7] B.D. Sleeman. On diffraction at short wave lengths by a prolate spheroid. J. Inst. Maths. Applics. **5** (1969) 432-442.

[8] B.D. Sleeman. Towards the validity of the geometrical theory of diffraction. Proc. Conf. on Function Theoretic Methods in Partial Differential Equations. Lecture notes in Mathematics, Vol. 561, Springer-Verlag (1976) 443-458.

[9] F. Ursell. On the short-wave asyptotic theory of the wave equation $(\nabla^2 + k^2)\phi = 0$. Proc. Camb. Philos. Soc. **53** (1957) 115-133.

[10] F. Ursell. On the rigorous foundation of short-wave asymptotics. Proc. Camb. Philos. Soc. **62** (1966) 227-244.

[11] F. Ursell. Creeping modes in a shadow. Proc. Camb. Philos. Soc. **64** 171-191.

[12] G.N. Watson. The diffraction of electric waves by the earth. Proc. Roy. Soc. London Ser.A **95** (1918), 83-99.

[13] T.T. Wu and S.R. Sheshadri. The electromagnetic theory of light II. Harvard University, Cruft Laboratory, Scientific Report No. 22 (1958).

Wave propagation in thin plates of elastic mixtures

C. CONSTANDA

1. INTRODUCTION

It is known from the classical linear theory (see, for example, [1]) that plane elastic waves do not propagate in a thin plate in the same way they do in an infinite medium. For the latter there are two types of such waves: longitudinal, when the displacement is in the direction of propagation, and transverse, for which the displacement occurs in a plane perpendicular to the direction of propagation.

If a thin elastic plate executes only extensional motions, then there are two so-called longitudinal waves propagating through it, both in the plane of the plate and generating displacements perpendicular to each other; in the case of bending, the resulting flexural waves propagate in a direction normal to the plane of the plate and their velocity is proportional to the wave number.

As is shown in [2], if the infinite medium is a mixture of two elastic materials, then for both longitudinal and transverse waves the wave number is in general complex, which means that the waves are damped. Using the theory developed in [3], in this paper we investigate the propagation of plane waves in a thin plate consisting of a mixture of two elastic solids and show that while for longitudinal waves the relation between the infinite mixture and the plate is similar to that in the classical theory, flexural waves are more complex here and their number depends on frequency and the magnitude of the elastic constants.

2. THE EQUATIONS OF PLATE THEORY

Throughout this paper Greek indices take the values 1,2, a superposed dot denotes differentiation with respect to time, $(\cdots)_{,\alpha} = \partial(\cdots)/\partial x_\alpha$, and the convention of summation over repeated indices is understood.

We consider an isotropic mixture of two homogeneous elastic solids occupying a three-dimensional domain B bounded by the planes $x_3 = \pm h/2$ (h = constant).

The equations for extensional motions of the plate in the absence of body forces and of forces acting on the surface are [3]

$$\left[\frac{E_1}{2(1-\nu_1)} + \lambda_5\right]\omega_{\beta,\beta\alpha} + \left[\frac{E_1}{2(1+\nu_1)} - \lambda_5\right]\omega_{\alpha,\beta\beta} + \left[\frac{E_3}{2(1-\nu_3)} - \lambda_5\right]\eta_{\beta,\beta\alpha}$$
$$+ \left[\frac{E_3}{2(1+\nu_3)} + \lambda_5\right]\eta_{\alpha,\beta\beta} = \rho_1\ddot{\omega}_\alpha + \alpha(\dot{\omega}_\alpha - \dot{\eta}_\alpha),$$
$$\left[\frac{E_3}{2(1-\nu_3)} - \lambda_5\right]\omega_{\beta,\beta\alpha} + \left[\frac{E_3}{2(1+\nu_3)} + \lambda_5\right]\omega_{\alpha,\beta\beta} + \left[\frac{E_2}{2(1-\nu_2)} + \lambda_5\right]\eta_{\beta,\beta\alpha}$$
$$+ \left[\frac{E_2}{2(1+\nu_2)} - \lambda_5\right]\eta_{\alpha,\beta\beta} = \rho_2\ddot{\eta}_\alpha - \alpha(\dot{\omega}_\alpha - \dot{\eta}_\alpha), \qquad (2.1)$$

where $\omega_\alpha(x_1,x_2,t), \eta_\alpha(x_1,x_2,t)$ are the displacements in the two constituents of the mixture, ρ_1, ρ_2 their densities, and E_i, ν_i ($i = 1,2,3$), α, λ_5 elastic constants.

In the case of bending, in the absence of body forces and of forces and moments acting on the faces, and disregarding rotatory inertia in both constituents of the mixture and the "mixed" transverse shear effect, the equations of motion are [3]

$$(D_1\omega_3 + D_3\eta_3)_{,\alpha\alpha\beta\beta} + 4h\lambda_5(\omega_3 - \eta_3)_{,\alpha\alpha} + h\rho_1\ddot{\omega}_3$$
$$- \alpha I(\dot{\omega}_3 - \dot{\eta}_3)_{,\alpha\alpha} + \alpha(\dot{\omega}_3 - \dot{\eta}_3) = 0,$$
$$(D_3\omega_3 + D_2\eta_3)_{,\alpha\alpha\beta\beta} - 4h\lambda_5(\omega_3 - \eta_3)_{,\alpha\alpha} + h\rho_2\ddot{\eta}_3$$
$$+ \alpha I(\dot{\omega}_3 - \dot{\eta}_3)_{,\alpha\alpha} - \alpha(\dot{\omega}_3 - \dot{\eta}_3) = 0, \qquad (2.2)$$

where $\omega_3(x_1,x_2,t)$, $\eta_3(x_1,x_2,t)$ are the deflections, $I = h^3/12$, and $D_i = E_i I/(1-\nu_i^2)$ ($i = 1,2,3$) rigidity coefficients similar to that in the classical theory.

3. LONGITUDINAL WAVES

For small amplitude harmonic plane waves propagating in the x_1-direction we replace ω_α and η_α in (2.1) by

$$\omega_\alpha = a_\alpha \exp\{i(kx_1-\Omega t)\},\ \eta_\alpha = b_\alpha \exp\{i(kx_1-\Omega t)\},$$

where a_α, b_α are constants, k is the wave number, and Ω the frequency. It then follows that (2.1) are satisfied if

$$\begin{aligned}(p_1k^2-\rho_1\Omega^2-\alpha\Omega i)a_1+(p_3k^2+\alpha\Omega i)b_1 &= 0,\\ (p_3k^2+\alpha\Omega i)a_1+(p_2k^2-\rho_2\Omega^2-\alpha\Omega i)b_1 &= 0,\end{aligned} \tag{3.1}$$

and

$$\begin{aligned}(q_1k^2-\rho_1\Omega^2-\alpha\Omega i)a_2+(q_3k^2+\alpha\Omega i)b_2 &= 0,\\ (q_3k^2+\alpha\Omega i)a_2+(q_2k^2-\rho_2\Omega-\alpha\Omega i)b_2 &= 0,\end{aligned} \tag{3.2}$$

where

$$p_i = E_i/(1-\nu_i^2)\ (i = 1,2,3),\ q_\alpha = E_\alpha/2(1+\nu_\alpha)-\lambda_5,$$
$$q_3 = E_3/2(1+\nu_3)+\lambda_5 .$$

The system (3.1) has non-zero solutions if and only if

$$(p_1p_2-p_3^2)k^4-[(\rho_2p_1+\rho_1p_2)\Omega^2+(p_1+p_2+2p_3)\alpha\Omega i]k^2+\rho_1\rho_2\Omega^4+\rho\Omega^3\alpha i=0 \tag{3.3}$$

where $\rho = \rho_1+\rho_2$.

If the diffusive force [2] is large, that is, α is large, and if $p_1p_2 \neq p_3^2$, then the solutions of (3.3) are

$$k^2 = \frac{(p_1+p_2+2p_3)\Omega\alpha}{p_1p_2-p_3^2}\, i\ , \tag{3.4}$$

$$k^2 = \frac{\rho\Omega^2}{p_1+p_2+2p_3} + \frac{\Omega[\rho_1(p_2+p_3)-\rho_2(p_1+p_3)]^2}{(p_1+p_2+2p_3)^2\rho\alpha}\, i\ . \tag{3.5}$$

The wave (3.4) has slow velocity and is heavily damped, whereas (3.5) is lightly damped, so that at some distance inside the plate it becomes the predominant wave. Its velocity does not depend on frequency and is equal to that corresponding to a single solid of density ρ and for which $4(\lambda+\mu)/(\lambda+2\mu) = p_1+p_2+2p_3$.

If $p_1p_2 = p_3^2$, then (3.3) yields

$$k^2 = \frac{\Omega^2}{\Omega^2(\rho_1 p_2+\rho_2 p_1)^2+\alpha^2(p_1+p_2+2p_3)^2}\left[\Omega^2\rho_1\rho_2(\rho_1 p_2+\rho_2 p_1)\right.$$

$$\left.+\rho\alpha^2(p_1+p_2+2p_3)+\alpha\Omega(\rho_1^2 p_2+\rho_2^2 p_1-2\rho_1\rho_2 p_3)\alpha\Omega i\right], \qquad (3.6)$$

and for α large we recover (3.5).

If the diffusive force is negligible, then we ignore the terms in α in (3.3) and obtain

$$k^2 = \frac{\Omega^2}{2(p_1 p_2-p_3^2)}\{\rho_1 p_2+\rho_2 p_1\pm[(\rho_1 p_2-\rho_2 p_1)^2+4\rho_1\rho_2 p_3^2]^{\frac{1}{2}}\}, \qquad (3.7)$$

provided $p_1 p_2 \neq p_3^2$. The values (3.7) are always real. If $p_1 p_2 > p_3^2$ and $\rho_1 p_2 + \rho_2 p_1 > 0$, we then obtain two undamped waves propagating with velocities independent of Ω. If $p_1 p_2 > p_3^2$ but $\rho_1 p_2 + \rho_2 p_1 = 0$, then only one such wave propagates.

If $p_1 p_2 = p_3^2$, from (3.6) we obtain

$$k^2 = \frac{\Omega^2\rho_1\rho_2}{\rho_1 p_2+\rho_2 p_1} .$$

When there is no interaction between the two constituents, we have [3] $p_\alpha = 4\mu_\alpha(\lambda_\alpha+\mu_\alpha)/(\lambda_\alpha+2\mu_\alpha)$, $p_3 = 0$, and (3.6) reduces to the classical solutions for the two individual solids [1]

$$k_\alpha^2 = \frac{\rho_\alpha\Omega^2(\lambda_\alpha+2\mu_\alpha)}{4(\lambda_\alpha+\mu_\alpha)} .$$

The system (3.2) has non-zero solutions if and only if

$$(q_1 q_2-q_3^2)k^4-[(\rho_2 q_1+\rho_1 q_2)\Omega^2+(q_1+q_2+2q_3)\Omega\alpha i]k^2+\rho_1\rho_2\Omega^4+\rho\Omega^3\alpha i = 0.$$

As in the classical theory, this equation coincides with that which yields the values of k for transverse waves in an infinite mixture of two solids [2].

4. FLEXURAL WAVES

Replacing

$$\omega_3 = a\exp\{i(kx_1-\Omega t)\}, \quad \eta_3 = b\exp\{i(kx_1-\Omega t)\},$$

with a,b constants, in (2.2) and reasoning as in the preceding section, we obtain that the wave number k satisfies the equation

$$(D_1D_2-D_3^2)k^8-(D_1+D_2+2D_3)(4h\lambda_5+\alpha I\Omega i)k^6$$

$$-[(\rho_2D_1+\rho_1D_2)\Omega+(D_1+D_2+2D_3)\alpha i]h\Omega k^4$$

$$+(4h\lambda_5+\alpha I\Omega i)h\rho\Omega^2k^2+(\rho_1\rho_2\Omega+\rho\alpha i)h^2\Omega^3 = 0.$$

For simplicity we examine the case when the only interaction terms in the partial stresses are due to volume changes. Then [3] $\lambda_5 = D_3 = 0$ and the above equation takes the form

$$D_1D_2k^8-(D_1+D_2)\alpha I\Omega ik^6-[(\rho_2D_1+\rho_1D_2)\Omega+(D_1+D_2)\alpha i]h\Omega k^4$$

$$+ \alpha Ih\rho\Omega^3ik^2+(\rho_1\rho_2\Omega+\rho\alpha i)h^2\Omega^3 = 0. \tag{4.1}$$

When the diffusive force is small, that is, when α is small, (4.1) yields

$$k_\gamma^2 = c_\gamma\Omega + \frac{i}{2}\left(\frac{I\Omega}{D_\gamma} + \frac{c_\gamma}{\rho_\gamma}\right)\alpha, \tag{4.2}$$

$$k_\gamma^2 = -c_\gamma\Omega + \frac{i}{2}\left(\frac{I\Omega}{D_\gamma} - \frac{c_\gamma}{\rho_\gamma}\right)\alpha, \tag{4.3}$$

where $c_\gamma = (h\rho_\gamma/D_\gamma)^{\frac{1}{2}}$, provided that $p_2D_1 \neq \rho_1D_2$.

The waves (4.2) propagate with velocities $(\Omega/c_\gamma)^{\frac{1}{2}}$ and are lightly damped. The waves (4.3) propagate only if $\Omega > 12/c_\gamma h^2$. They travel faster than (4.2) and are more heavily damped.

If there is no interaction between the two constituents of the mixture, we have $1/c_\gamma^2 = E_\gamma h^2/12\rho_\gamma(1-\sigma_\gamma^2)$, $\alpha = 0$. In this case (4.3) do not propagate, while (4.2) reduce to the classical flexural waves [1] through the two individual solids.

If $\rho_2D_1 = \rho_1D_2$, then we obtain from (4.1)

$$k^2 = c\Omega, \tag{4.4}$$

$$k^2 = c\Omega + \frac{i}{2}\left(\frac{1}{D_1} + \frac{1}{D_2}\right)\left(\Omega I + \frac{h}{c}\right)\alpha, \tag{4.5}$$

$$k^2 = -c\Omega + \frac{i}{2}\left(\frac{1}{D_1} + \frac{1}{D_2}\right)\left(\Omega I - \frac{h}{c}\right)\alpha, \tag{4.6}$$

where $c^2 = \rho h/(D_1+D_2)$.

The wave (4.4) travels through both constituents as if they were one solid of density ρ and rigidity coefficient $D_1 + D_2$, and its velocity $(\Omega/c)^{\frac{1}{2}}$ is the same for either solid. The wave (4.5) travels with the same velocity as (4.4) and is lightly damped.

The wave (4.6) propagates only if $\Omega > 12/ch^2$. As $\Omega \to 12/ch^2$, it travels much faster than (4.5) and is more heavily damped, so that at some distance inside the plate (4.5) will predominate. As $\Omega \to \infty$, the ratio of the velocities of (4.5) and (4.6) tends to $\frac{1}{4c}\left(\frac{1}{D_1} + \frac{1}{D_2}\right)\alpha$.

If the diffusive force is large, that is, if α is large, (4.1) yields

$$k^2 = iI\Omega\left(\frac{1}{D_1} + \frac{1}{D_2}\right)\alpha, \tag{4.7}$$

$$k^2 = \frac{1}{p}\left[-1+i\,\frac{D_1D_2}{pI\Omega(D_1+D_2)\alpha}\,\frac{(p^2c_1^2\Omega^2-1)(p^2c_2^2\Omega^2-1)}{p^2c^2\Omega^2-1}\right], \tag{4.8}$$

$$k^2=c\Omega\left[1-i\,\frac{D_1D_2\Omega}{h^2\rho\alpha}\,\frac{(c^2-c_1^2)(c^2-c_2^2)}{pc\Omega+1}\right], \tag{4.9}$$

$$k^2 = -c\Omega\left[1+i\,\frac{D_1D_2\Omega}{h^2\rho\alpha}\,\frac{(c^2-c_1^2)(c^2-c_2^2)}{pc\Omega-1}\right], \tag{4.10}$$

where $p = h^2/12$, provided that $\Omega \neq (pc)^{-1}$.

Suppose for convenience that $c_1 \leqslant c_2$. Then we have $c_1 \leqslant c \leqslant c_2$.

The wave (4.7) is slow and heavily damped. The wave (4.8) propagates only if $c_1 \neq c_2$ and either $1/pc_2 < \Omega < 1/pc$ or $\Omega > 1/pc_1$. In either case it is faster than (4.7) and its damping coefficient is independent of Ω and proportional to h^{-1}.

If $c_1 \neq c_2$, the wave (4.9) tends to behave like a wave propagating through a single solid with velocity $(c\Omega)^{\frac{1}{2}}$ and suffering little damping. If $c_1 = c_2$, the velocity is the same but there is no damping.

The wave (4.10) propagates only if $c_1 \neq c_2$ and $\Omega > 1/pc$. As $\Omega \to \infty$, it travels much faster than (4.8) and is more heavily damped.

If $\Omega = (pc)^{-1}$, then

$$k^2 = i\,\frac{h}{c}\left(\frac{1}{D_1} + \frac{1}{D_2}\right)\alpha,$$

$$k^2 = \frac{1}{p}\left[1-i\,\frac{D_1D_2\Omega}{2h^2\rho pc\alpha}(c^2-c_1^2)(c^2-c_2^2)\right].$$

REFERENCES

1. L. D. Landau, E. M. Lifshitz; Course of Theoretical Physics, vol. 7: Theory of Elasticity, 2nd ed., Oxford - London - Edinburgh - New York - Toronto - Sydney - Paris - Braunschweig, Pergamon Press, 1970.
2. T. R. Steel; Applications of a theory of interacting continua, Quart. J. Mech. Appl. Math., 20, 57-72, 1967.
3. C. Constanda; Bending of thin plates in mixture theory, Acta Mech., 40, 109-115, 1981.

Dr. C. Constanda,
Department of Mathematics,
University of Strathclyde,
Glasgow, Scotland.

The biharmonic equation and its finite difference approximations

R.F.McLEAN

1. INTRODUCTION

The determination of the natural frequencies and associated mode shapes of vibrating plates has intrigued mathematicians and engineers for almost two centuries. Published work is in evidence as far back as 1802, when the experimenter Chladni [1] produced his notable work. Since then attention to this problem was given in turn by Rayleigh [2], Voight [3] and Love [4] without providing an adequate solution suitable for engineering applications. The most recent work on this topic is due to Warburton [5] who provides the greatest insight into this important engineering problem. His work, based on the Rayleigh-Ritz approach, gave a much needed practical method for the engineering profession. However, only a limited number of plate shapes can be tackled in this way and the author considered the biharmonic equation and its finite difference approximations as an approach to plate vibration. A number of difficulties arose, particularly in the handling of boundary conditions. In the engineering profession, it is not acceptable to state that a solution is not possible; reasonable assumptions must be made to provide some form of solution. Such assumptions are made in this paper together with their justifications and limitations.

2.1 The Basic Equations

The governing differential equation for the free vibration of thin flat plates is the biharmonic equation with an added forcing term due to the inertia of the plate. As pointed out by Love [4] and Timoshenko [5] there is no complete analytical solution to this problem and, except for the special cases where the exact solutions exist, approximate methods must be used.

To apply the finite difference technique to the solution of this differential equation, the plate is sub-divided by a series of horizontal and vertical mesh lines. The biharmonic equation is then replaced by an equivalent finite difference equation which is considered to hold at the intersection of the mesh lines. A series of algebraic equations is then built up by considering each of the points on the plate in turn, the relative amplitude of vibration of the points around the mesh point being given in terms of a non-dimensional frequency factor by these equations.

2.2 Biharmonic Operator for a Square Lattice

The governing biharmonic equation for the vibration of flat plates with the forcing term on the right hand side is,

$$\frac{\partial^4 w}{\partial x^4} + 2\frac{\partial^4 w}{\partial x^2 \partial y^2} + \frac{\partial^4 w}{\partial y^4} = -\frac{12\rho(1-\nu^2)}{Eg\, t^2}\frac{\partial^2 w}{\partial t^2}$$

where $w = f(x, y, t)$. With the assumption that the vibration of the plate is sinusoidal, it follows that,

$$\frac{\partial^4 w}{\partial x^4} + 2\frac{\partial^4 w}{\partial x^2 \partial y^2} + \frac{\partial^4 w}{\partial y^4} = \lambda w \qquad (2.1)$$

where $\lambda = A\omega^2 \sin\omega t$

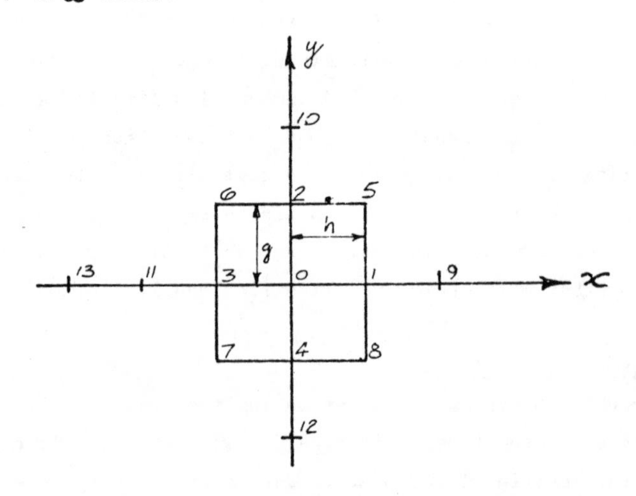

FIGURE 2.1

Referring to Figure 2.1, it can be seen that for central differences in the x-direction,

$$\left.\begin{aligned} (h^4 D^4)_0 &= w_{11} - 4w_3 + 6w_0 - 4w_1 + w_9 \\ & \qquad\qquad\qquad\qquad e = 0(h^2) \\ (h^2 D^2)_0 &= w_3 - 2w_0 + w_1 \end{aligned}\right\} \quad (2.2)$$

substituting equations (2.2) into equation (2.1) and re-arranging gives,

$$20w_0 - 8\left[w_1 + w_2 + w_3 + w_4\right] + 2\left[w_5 + w_6 + w_7 + w_8\right] + \left[w_9 + w_{10} + w_{11} + w_{12}\right] = h^4 \lambda w_0 \quad (2.3)$$

In diagrammatic form this can be shown in Figure 2.2 as:

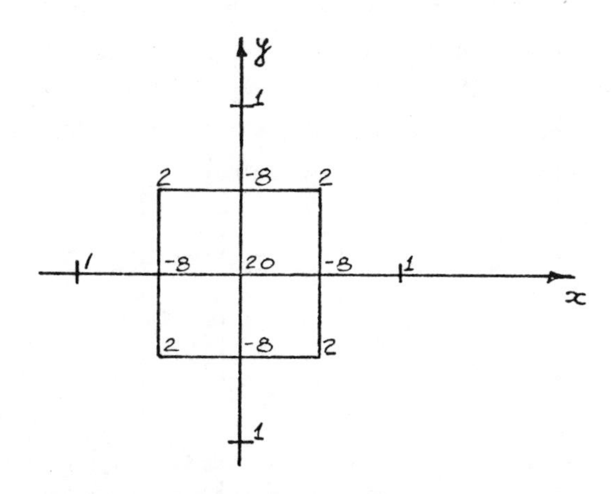

FIGURE 2.2

The method of Bickley [6] can be used as an alternative to derive the above and more accurate 'molecule' diagrams, by applying a Taylor series expansion in terms of central differences.

The above equation (2.3) together with an extension to include points more widely spaced than those chosen in Figure 2.1 can be used to evaluate molecule diagrams of varying accuracy and complexity. The diagrams are obtained by solving the equations as a set of simultaneous equations and eliminating the differential terms of the right hand sides. Elimination of terms below h^6 leads to equation (2.3). More accurate expressions can be evaluated and are comprehensively listed in Collatz [8]. However, Mathur [9], has shown that by a correction procedure, a low order accuracy molecule diagram, such as shown in Figure 2.2, can be used to give an accuracy comparable to that expected only with a higher order, more complex, molecule diagram.

For rectangular plates, it is not always possible to fit a square mesh over the plate with mesh lines coinciding with the geometric boundaries of the plate. To overcome this difficulty, two methods were tried. Firstly, a square mesh was used and a series of irregular finite difference operators were constructed, taking into account the difference in length of the mesh lines within a lattice adjacent to the boundary. The second method considered the mesh lines to coincide with the geometric boundaries of plate by considering a rectangular rather than a square mesh. The latter method gave more accurate results, so this procedure was adopted and the finite difference operators were developed for a rectangular mesh.

2.3 Biharmonic Operator for a Rectangular Mesh

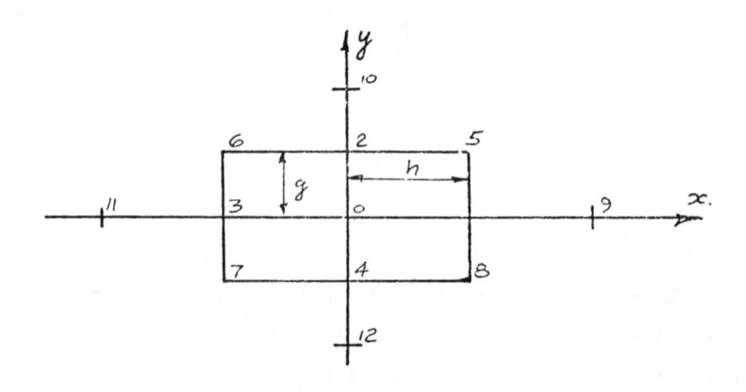

FIGURE 2.3

As for the square mesh, the derivatives $\frac{\partial^4 w}{\partial x^4}$ and $\frac{\partial^4 w}{\partial y^4}$ remain unaltered apart from the change of interval length, then applying the biharmonic to point 0 of Figure 2.3 gives,

$$\left[8R + 6(S+1)\right] w_0 - \left[4(R+1)\right] \cdot (w_1 + w_3) - \left[4(R+S)\right] \cdot (w_2 + w_4) + \left[2R\right] \cdot (w_5 + w_6 + w_7 + w_8) + \left[1\right] \cdot (w_9 + w_{11}) + \left[S\right] \cdot (w_{10} + w_{12}) = \lambda h^4 w_0 \qquad (2.4)$$

with $R = \dfrac{h^2}{g^2}$ and $S = \dfrac{h^4}{g^4}$

which can be expressed diagrammatically as:

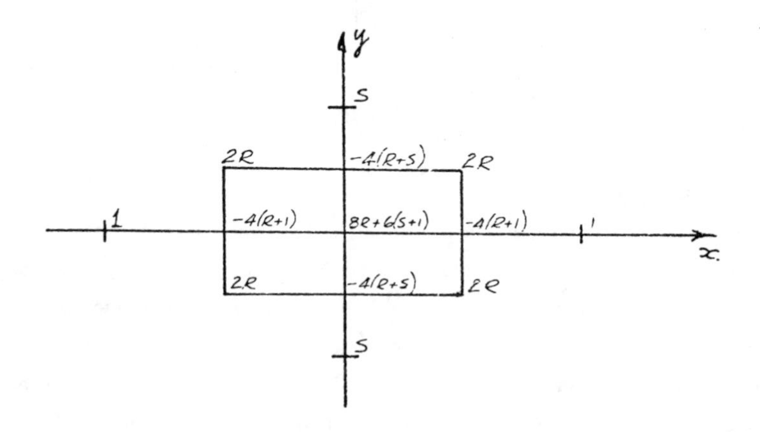

FIGURE 2.4

It is convenient when handling corner points of free boundary plates to have a modified biharmonic operator which excludes one of the corner points (e.g. point 5) shown in Figure 2.3. This can be achieved by using backward differences for the $\dfrac{\partial^2 w}{\partial y^2}$ term. The backward difference is constructed to the same order of truncation error as the equivalent central difference operator to enable the accuracy of the ensuing molecule diagram to be substantially unaltered.

The modified biharmonic operator is then constructed to give:

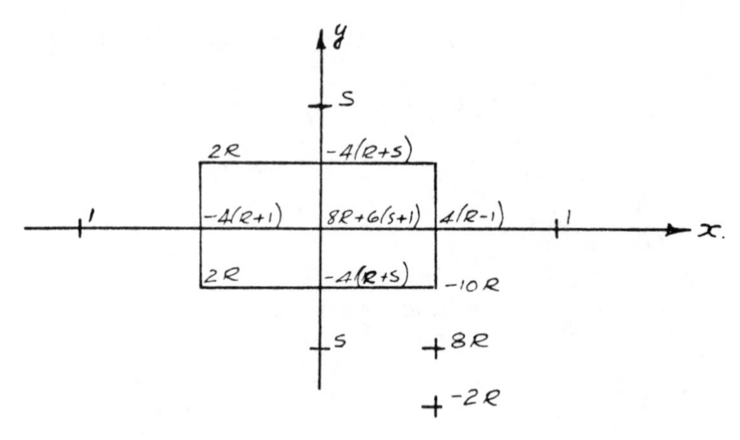

FIGURE 2.5

2.4 Boundary Conditions

The molecule diagrams formulated in the previous section apply at the intersection of an infinite mesh. For a given plate shape the mesh must terminate at the boundaries and consequently, a molecule diagram for a mesh point near, or at a boundary, will overflow the edge of the plate. Finite difference equations are therefore built up to satisfy the boundary conditions and these equations are then used to express the relationship between points lying outside the boundary to points lying on, or within, the plate.

2.4.1 Encastre edge condition

The amplitude of vibration of mesh points lying on an encastre edge is zero and does not require to be considered as a mesh point for the formulation of the set of algebraic equations. The coefficient for this point will therefore be zero when considering it as part of the molecule diagram for other mesh points.

From the finite difference equation, it can be shown that a point lying one mesh length outside an encastre boundary can be replaced by the point which is its mirror image lying one mesh length inside the boundary.

2.4.2 Simply supported edge condition

Again, a point situated in a simply supported boundary need not be included as a mesh point and only internal mesh points need be considered for this boundary condition. A mesh point situated one mesh length from a simply supported edge can be referred to its mirror image point lying within the boundary of the plate but with a negative sign affixed to it.

2.4.3 Free edge condition

For this boundary condition, the shear force and bending moment normal to the edge of the plate are both equal to zero and, since the amplitude at this edge is not zero, molecule diagrams must be constructed for points lying on the boundary. Since the boundary conditions include third order derivatives, they involve considerably more labour than the previous two conditions.

On the free edge the boundary conditions are,

$$\left[\frac{\partial^2 w}{\partial y^2} + \nu \frac{\partial^2 w}{\partial x^2}\right]_{y=0} = 0$$

and

$$\left[\frac{\partial^3 w}{\partial y^3} + (2 - \nu) \frac{\partial^3 w}{\partial y . \partial x^2}\right]_{y=0} = 0$$

To apply the biharmonic operator to point (0), the points (6), (2), (5) and (10) must be referred to points lying within the plate. This can be achieved by using the first of the boundary equations to reflect, by central differences, the points lying one mesh length from the boundary to the truncation error $0(h^2)$. The second of the boundary equations involves the third derivative. Using central differences for the point (0), to the same order of accuracy as in the previous case, the point (10) can be expressed in terms of points within the confines of the plate together with point (2) lying outwith the plate. The first boundary condition is then used to reflect point (2) as before, i.e. for the first boundary condition,

$$w_2 = 2w_0 - w_4 - \frac{\nu}{R}\left[w_3 - 2w_0 + w_1\right]$$

Similarly the points (5) and (6) can be represented by a corresponding distribution of points.

For the second boundary condition, using a contribution of central and backward differences, the point lying two mesh lengths from the free edge can be found to be,

$$w_{10} = \left[4\left(1 + \frac{\nu}{R}\right) + \frac{6}{R}(2 - \nu)\right] \cdot w_0 - \left[2\frac{\nu}{R} + 3(2 - \nu)/R\right] \cdot$$
$$(w_1 + w_3) - \left[4 + 8(2 - \nu)/R\right] w_4 + \left[4(2 - \nu)/R\right] \cdot (w_7 + w_8)$$
$$+ \left[\frac{2}{R}(2 - \nu) + 1\right] \cdot w_{12} - \left[\frac{(2 - \nu)}{R}\right] \cdot (w_{13} + w_{14}) \qquad (2.5)$$

Equations (2.4) and (2.5) can be used to reflect points (2), (5), (6) and (10), to give the biharmonic operator for a point lying on a free boundary as,

$$\left[6 + 25 - 10\nu R - 12(R - \nu)\right] \cdot w_0 - \left[R(6 - 5\nu) + 4(1 - 2\nu)\right] \cdot$$
$$(w_1 + w_3) - \left[4R(4 - 2\nu + R)\right] \cdot w_4 + \left[4R(2 - \nu)\right] \cdot (w_7 + w_8)$$
$$+ \left[1 - 2\nu\right] \cdot (w_9 + w_{11}) + \left[2R(2 - \nu) + 2S\right] \cdot w_{12} + \left[R(2 - \nu)\right] \cdot$$
$$(w_{13} + w_{14}) = 0 \qquad (2.6)$$

2.4.4 Corner points

To construct molecule diagrams at the intersection of two boundaries, reflection of points lying outwith the confines of the plate are carried out using the molecule diagram relevant to the boundary across which the point is being reflected. Apart from the condition of a free/free intersection, the molecule diagrams can be built up using the equations derived previously. To derive the molecule diagram for a point occurring at the intersection of a free/free boundary the modified biharmonic Figure 2.5 is used to eliminate point (5). In order to reflect points (1) and (2), without double-reflection, new reflection diagrams must be obtained for these points using a combination of central and backward differences.

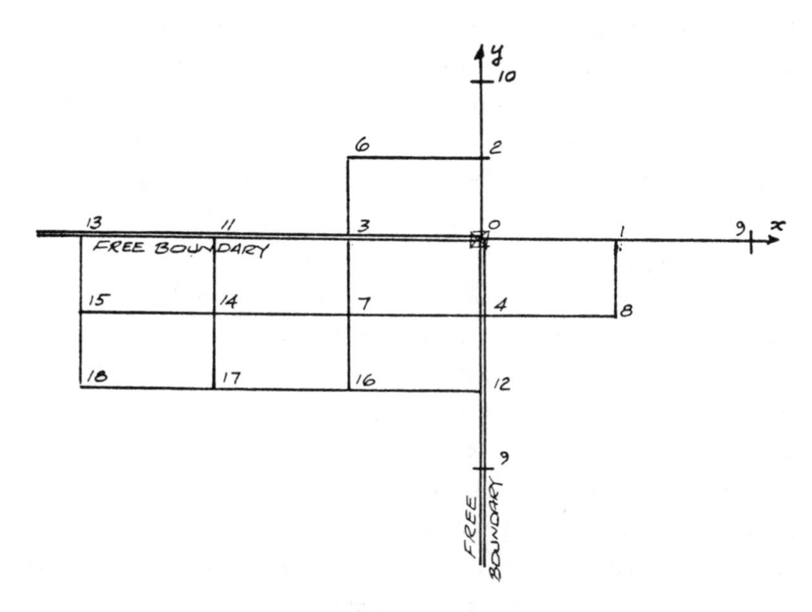

FIGURE 2.6

Applying the boundary conditions of Section 2.4.3 to Figure 2.6 gives, for the first condition,

$$w_2 = 2w_0 - w_4 - \frac{\nu}{R}\left[2w_0 - 5w_3 + 4w_{11} - w_{13}\right] \tag{2.7}$$

and for the second boundary condition,

$$\begin{aligned} w_{10} = &\left[4\left(1 - \frac{\nu}{R}\right) - \frac{6}{R}(2-\nu)\right]\cdot w_0 + \left[\left(\frac{10\,\nu}{R}\right) + \frac{15}{R}(2-\nu)\right]\cdot w_3 \\ &- \left[4 - \frac{8}{R}(2-\nu)\right]\cdot w_4 - \left[\frac{20}{R}(2-\nu)\right]\cdot w_7 - \left[\frac{8\,\nu}{R} + \frac{12}{R}(2-\nu)\right]\cdot w_{11} \\ &+ \left[1 - \frac{2}{R}(2-\nu)\right]\cdot w_{12} + \left[\frac{2\,\nu}{R} + \frac{3}{R}(2-\nu)\right]\cdot w_{13} + \left[\frac{16}{R}(2-\nu)\right]\cdot w_{14} \\ &- \left[\frac{4}{R}(2-\nu)\right]\cdot w_{15} + \left[\frac{5}{R}(2-\nu)\right]\cdot w_{16} - \left[\frac{4}{R}(2-\nu)\right]\cdot w_{17} \\ &+ \left[\frac{1}{R}(2-\nu)\right]\cdot w_{18} \end{aligned} \tag{2.8}$$

2.4.5 Hole in plate

As in the previous case, this involves an intersection of a free/ free boundary condition. However, since there are less points lying outwith the plate, it is much easier to handle.

2.4.6 Partial boundary condition

The molecule diagrams developed along a boundary relate the

The molecule diagrams developed along a boundary relate the amplitude of vibration of specific points as a function of the amplitude of the points surrounding it. Since the points are chosen sequentially along the edge of the plate, a change in boundary condition along any particular plate edge can be accommodated.

2.4.7 Mixed boundaries

In many engineering applications, a plate edge will not fall into one of the clear cut boundary conditions but may, for example, be somewhere between the encastre and simply supported condition. If such boundaries are expressed in the proportion of the three main boundary conditions, they can be handled by super-imposing the expressions derived in this section.

2.4.8 Skewed and curved boundaries

No new techniques are required for the case of skewed and curved boundaries other than the transformation of the boundary conditions which are known along and normal to the boundary, to equivalent conditions lying in direction of the mesh lines.

3. SOLVING THE EQUATIONS

At the commencement of this project a computer solution to the finite difference equations was available by an iterative procedure. This method was used to check the initial feasibility of using the finite difference technique in the solution of the free vibration of flat plates. It found the eigenvalues in descending order with each root depending upon the accuracy of the previously found higher value; the lower values were therefore the least accurate. Since the engineer is more interested in the values for the lower natural frequencies, this procedure was not acceptable, unless the matrix was first of all inverted, to give the reciprocal of the lowest natural frequency as the highest eigenvalue. However, the inversion procedure introduces rounding errors and the accuracy of succeeding higher values become progressively less accurate. This, coupled with the fact that the programme found difficulty in discerning between discrete mode shapes which had equal frequencies made the iterative procedure, with inversion, unsatisfactory as a general solution.

An analysis technique was therefore developed to give a direct method of solution by similarity transformations capable of giving roots in the complex plane of unsymmetric matrices. Inverse iteration was then applied to obtain the eigenvector and hence the mode shape of the plate. This produced a more satisfactory solution and gave all roots to the same accuracy and handled the condition of discrete mode shapes having equal frequencies.

4.1 Computer Output and Results

A selection of nine plate shapes was considered to determine the suitability of finite difference techniques to the problem of the free vibration of flat plates.

The first three plates of the series had the same boundary condition on each of their four edges. Square plates were selected here to examine the facility of the method to give the coupled modes experienced predominantly in free and fixed square plates. The plate which had four of its edges simply supported was also used to consider the effect of varying the size and ratio of length to breadth of the chosen mesh.

The fourth plate shape was that of the square cantilever. Considerable difficulty was experienced when applying the methods of Section 2 to determine the torsional modes of vibration for this plate shape. As discussed in Section 6, a special technique had to be evolved to obtain a reasonable solution to the fundamental torsional mode of vibration for this plate.

The next two plate shapes to be considered were the cantilever and 'propped' cantilever conditions where the length to breadth ratio of the plate was 2:1. For the 'propped' cantilever condition the edge opposite the encastre boundary was considered to be simply supported. Experimental results were obtained for these plate shapes to enable a comparison to be made between the solutions obtained by finite difference and other methods.

In the remaining three plates, two opposite edges of a square plate were considered free while the other two edges were either wholly or partially, simply supported. The first plate of this series had two opposite edges simply supported and here the results could be compared with the energy method of Warburton together with the experimental results shown in Section 6. The next plate had similar boundary conditions to the above plate but had in addition, a square hole cut in the centre of the plate. The last plate was considered to have the simply supported boundary condition extending only over half the length of two of the boundaries, the remainder of these boundaries being free. Since the finite difference appeared to be the only method offering a solution to the last two plate shapes, an experimental check was made on these plates and appears in Section 6.

Use of symmetry was made for all plates with the exception of the plate having partially simply supported boundaries. The cantilever and 'propped' cantilever plates were considered to be either symmetric or asymmetric about the mid-line bisecting the encastre edge of the plate. The other bisected plates were considered to be either symmetric or asymmetric about the two mid-lines of the plates. This gave four conditions to be applied to the quarter plate being investigated. The first plate was further sub-divided by lines of symmetry joining opposite corners of the plate. However, to retain a more general approach, this was not used in the computer programme which compiled the matrices for any plate shape. The number of points considered within a section of the plate being examined was chosen to fit comfortably within the size of the digital computer available.

The computer programme calculated the eigenvalue of the matrix which is a function of frequency, mesh size and physical properties of the plate material.

Milne [9] suggests a method for the two dimension Laplacian operator in which, by means of a correction factor, a course relaxation pattern can be used to give the results which could only be considered obtainable from a much more accurate form of relaxation pattern. Mathur [8] extended this method to the biharmonic operator to give a correction factor of the form,

$$\sqrt{\lambda_c} = - 6 \log_e \left(1 - \frac{\sqrt{\lambda}}{6}\right)$$

where λ and λ_c are the eigenvalue and corrected eigenvalue of the matrix respectively.

4.2 Effect of Mesh Size

A square plate with four simply supported edges was used to investigate the effect of slight variations in the size and ratio of the length to breadth of the mesh. Figure 4.1 shows the results in terms of the natural frequency of a square, mild steel plate having the dimensions 10 in x 10 in x 0.25 in. A quarter of the plate is considered in this comparison and only the first three natural frequencies which correspond to modal shapes symmetric about both centre lines of the plate are calculated. It can be seen from this diagram that the variation in the fundamental frequency is less than 0.2% for any of the conditions chosen. From the higher mode shapes it can be seen that a reduction in mesh size from an 8 x 8 grid to a 7 x 7 grid introduces an error not greater than 0.3%. This small degree of error is maintained in the case of a rectangular mesh for the condition when the nodal line runs at right angles to the boundary which has the shortest mesh length. The error increases to 2% when the nodal lines run parallel to this boundary. This would be expected since by applying finite differences the mode shape in each direction is fitted to the number of mesh points in that direction. Consequently, by reducing the number of points in a direction in which the mode shape of the plate changes appreciably, such as the condition where a nodal line cuts this boundary, a reduction in accuracy would be expected.

MODE SHAPE	QUARTER PLATE SUBDIVIDED BY		MESH LENGTH RATIO	FREQ. c.p.s.
	No of HORIZONTAL MESH LINES	No of VERTICAL MESH LINES		
	8	8	1	478.75
	8	7	7/8	478.02
	8	6	3/4	477.65
	7	7	1	478.25
	8	8	1	2366.98
	8	7	7/8	2347.81
	8	6	3/4	2318.56
	7	7	1	2359.68
	8	8	1	2366.98
	8	7	7/8	2366.73
	8	6	3/4	2366.36
	7	7	1	2359.68

Variation in Mesh Dimensions
Square Plate With Four Simply Supported Edges
Plate Size 10 in x 10 in x 0.25 in

FIGURE 4.1

5. EXPERIMENTAL PROCEDURE

Four plate shapes were manufactured, the first two were cantilevers one of which was used to consider the effect of end inertia on the natural frequencies; the other to include the 'propped' cantilever condition. The third plate was machined as a square plate to investigate a combination of simply supported, partially simply supported and free edges. The fourth plate was similar to the previous plate, but had in addition a square hole machined from the centre.

Heiba [10] and other authors expressed doubts about the effectiveness of obtaining a fixed edge by clamping a uniform thickness plate at one end. To examine this condition experimentally, a plate was machined from a solid steel block to give a cantilever plate 8 in x 4 in x 0.25 in with a solid end mass. This end mass, initially 2.75 in thick, was reduced in half inch steps, until the end mass was the same thickness as the plate. At each reduction, the first six resonant frequencies were determined to ascertain the effect of end mass on natural frequency. When a change in frequency occurred, the plate was built up to its previous thickness by the addition of two quarter inch thick machined plates to correspond to the amount of metal previously removed. Finally, when the plate was reduced to uniform thickness, the end mass was built up to its original dimensions by a series of quarter inch thick plates.

During each of the above tests the holding down bolts were secured to large cast iron blocks by half inch bolts which were maintained at a uniform tension by the application of a torque wrench.

Figure 5.1 shows the results of these tests, the solid lines depicting the solid support and the dotted lines the effect of 'building up' by the quarter inch thick plates.

From these experiments it would appear that an encastre edge condition can be achieved by ensuring that the thickness of the end mass is at least five times the thickness of the plate material. It would also appear that lamination of a series of plates to form an end mass is not an effective method of obtaining the encastre edge condition.

6.1 Comparison of Results

Figures 6.1 to 6.9 give a tabulated comparison of the non-dimensional frequency factors for the finite difference and other methods. The frequency factor determined from Warburton's [5] equation (16) is multiplied by π^2 to conform to the non-dimensional frequency factor used by the other authors.

Significant points of discussion are made only for two selected plates in which the finite difference method requires careful consideration in its application. The first of these being the free/free plate and the second the cantilevered plate.

6.2 Square Plate With Four Edges Free

An initial set of results were obtained by applying the boundary

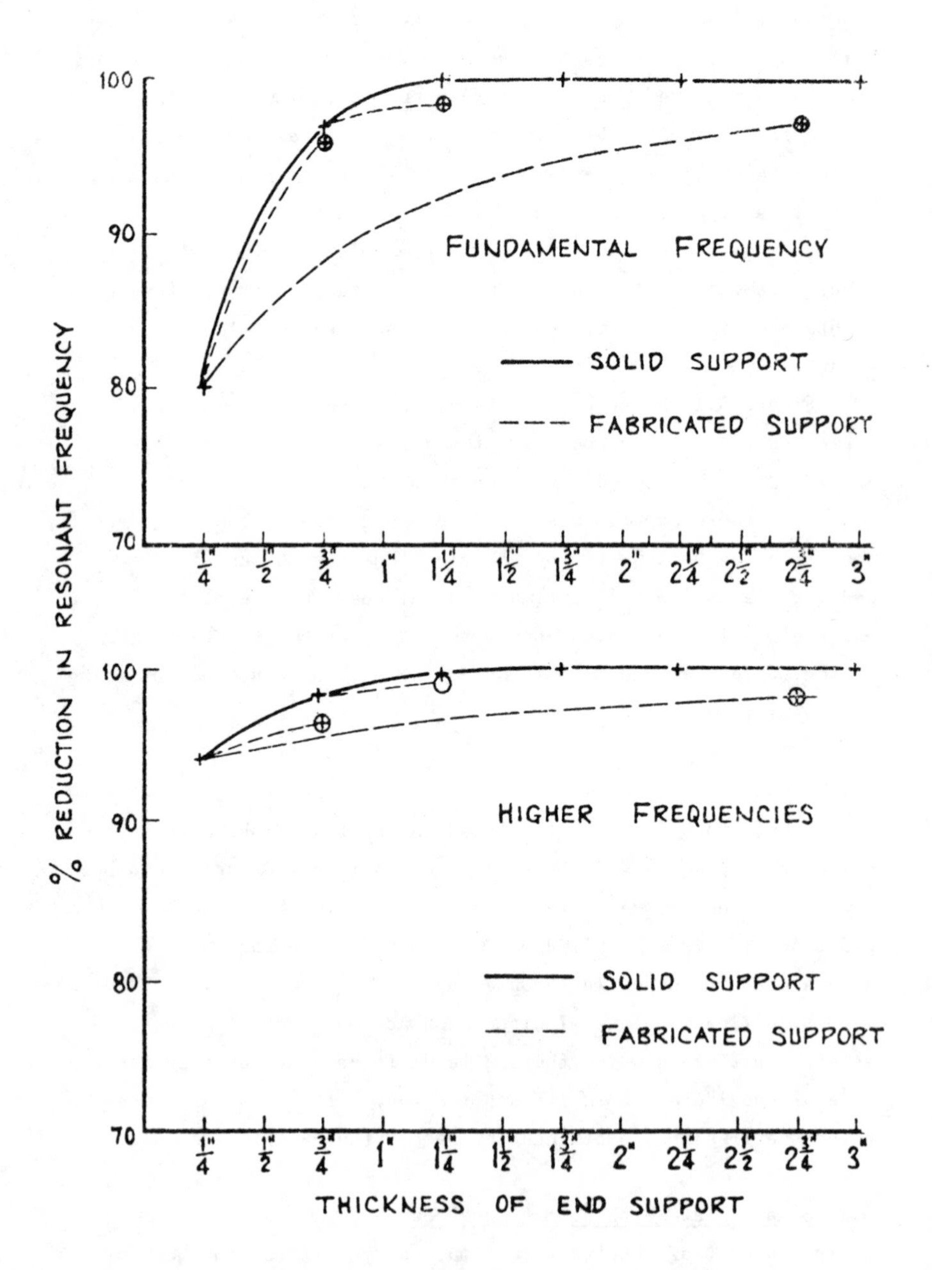

% REDUCTION IN RESONANT FREQUENCY
FUNDAMENTAL FREQUENCY
SOLID SUPPORT
FABRICATED SUPPORT
HIGHER FREQUENCIES
SOLID SUPPORT
FABRICATED SUPPORT
THICKNESS OF END SUPPORT

An initial set of results were obtained by applying the boundary conditions of Section 2.4.3 and the mode shapes are as would be expected from the discussion of this plate shape in Warburton's Paper. However, the frequencies are not to the accuracy expected. It was felt that the derivation of the unsymmetric molecule diagram for the corner point of a free/free plate could give this discrepancy. In Figure 2.6, the molecule diagram extends over three intervals in the horizontal direction and over four intervals in the vertical direction. As a simple expedient the unbalance in the two directions of the molecule diagram were interchanged and the mean of the two conditions used as a molecule diagram to give the modified results shown in Figure 6.3. The frequencies corresponding to this result are close to the frequency that would be expected from Warburton's Paper, but the mode shapes have deteriorated by applying this modification. However, since this modified molecule diagram did not conform to the rules applied to formulate finite difference equations, the mode shapes were not expected to be accurate. It would appear, therefore, that provided a symmetric molecule diagram can be evolved for the free/free corner point, the accuracy of the frequency and mode shape for this plate should be as good as that obtained for the previous two plate shapes.

6.3 Square Cantilever Plate

The results obtained for this plate shape were considerably less accurate than the results obtained for any other plate shape. The greatest discrepancy occurring in the torsional modes. The reason for these large discrepancies can be attributed to the boundary conditions for the encastre edge of the cantilever. The slope of the plate at this edge is considered to be zero in the direction normal to the encastre edge. This, together with the boundary condition of zero amplitude on this edge is generally sufficient to fully specify the encastre boundary condition. However, in the finite difference type of solution, this is not enough to attain an encastre edge. With this method, the amplitude of vibration of the mesh point being considered is related to the amplitude of the points in close proximity to it; all other points within the mesh having a coefficient equal to zero. When the boundary condition of zero amplitude is then applied to the encastre edge, the points on this boundary are eliminated since they have no amplitude. However, the elimination of these points is identical to considering that they do not exist as far as adjacent mesh points are concerned. The condition of zero amplitude can not therefore be achieved in this manner and by simply applying the boundary conditions of Section 2.4.1, the plate is merely constrained to have zero slope about this edge. This was illustrated by applying the boundary conditions of Section 2.4.1, for the fundamental torsional mode of the cantilever in which a nodal line extends from the centre of the encastre edge to the centre of the opposite free edge. For this condition, the plate merely rotated about the above nodal line at zero natural frequency, thereby executing the type of whole motion of a free plate. The additional boundary condition of zero slope along the encastre edge could not be applied to give the required rigidity along this edge.

One method tried to overcome this difficulty was to derive the biharmonic operator for all points along the encastre boundary. Since the amplitude of vibration must be equal along this boundary, the biharmonic operator for each of these points was super-imposed to give a common mesh point, chosen to be the mesh point at the centre of the boundary. The point was therefore related to all mesh points lying within a distance of two mesh lengths from the boundary. This has the effect of bringing the frequency away from zero but was still one tenth of the expected frequency for the fundamental torsional mode shape. This being due to the fact that the super-imposition of the molecule diagrams for the mesh points along the boundary specified only one mesh point on the boundary; the rest being excluded and therefore considered to have no effect. The encastre edge of the plate was then considered to extend for one mesh length along the free edges adjacent to the encastre boundary. Then using a partial boundary condition, a frequency of one third of that expected was obtained.

A similar result was obtained by considering the condition of zero slope to extend for one mesh along the fundamental torsional node line. It was felt that a combination of these two methods would have yielded a better solution but, they were not considered further since they violated the geometric boundary conditions of the plate and would not help towards giving a general solution.

The encastre boundary condition appeared to give trouble only when associated with three free edges to give the cantilevered plate. It was then decided that the fixing of one point on one of the free edges with respect to the boundary, might yield a solution. For the fundamental torsional mode shape a nodal line runs along the middle of the plate from the encastre to the opposite free edge. The point lying in the middle of this free edge was therefore considered to be equal to zero. A frequency approximately one third of the expected value was again obtained.

It then became apparent that the amplitude of the points on the encastre edge must be specified rather than being set to zero. Molecule diagrams were set up for the mesh points along this boundary and for the point located at the centre of the opposite free boundary. The amplitude of each of these points was considered to be equal but not necessarily equal to zero. This yielded a solution which was within 5% of the experimental solution obtained by Barton [11] for the fundamental torsional frequency. The other mode shapes given in Figure 6.4 are as derived from the application of the boundary conditions of Section 2.4.1.

The special techniques applied to the fundamental torsional mode of the square cantilever suggests that the cantilever, treated as a special case, could yield reasonable results.

NON-DIMENSIONAL FREQUENCY FACTORS

4 EDGES SIMPLY SUPPORTED

	Exact	Warburton	Finite Element	Finite Difference
	19.74	19.74	19.58	19.74
	49.36	49.34	-	49.22
	78.97	78.96	-	78.99
	98.72	98.70	97.49	97.74
	128.3	128.3	-	128
	167.8	167.8	-	164.3
	197.4	197.4	-	192.9

Figure 6.1

NON-DIMENSIONAL FREQUENCY FACTORS

4 EDGES - ENCASTRE

Warburton	Finite Element	Finite Difference
35.98	35.45	35.54
73.40	-	71.83
108.2	-	105.95
132.2	-	128.62
132.5	130.3	129.10
165.0	-	160.34
211.35	-	199.74
243.98	-	233.3
-	-	422.59

Figure 6.2

NON-DIMENSIONAL FREQUENCY FACTORS

4 EDGES FREE

Ritz	Warburton	Finite Difference
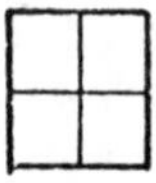 14.10	14.19 -	- 14.14
 20.55	22.38 19.67	17.02 19.11
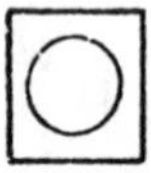 23.91	22.38 24.48	19.09 23.77
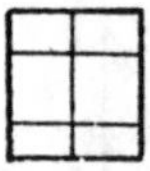 35.96	36.51 -	23.23 25.29
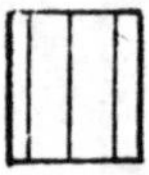 61.60	61.68 -	54.12 64.00
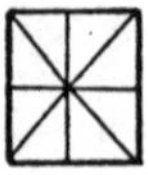 -	75.05 -	64.95 -

Figure 6.3

NON-DIMENSIONAL FREQUENCY FACTORS

SQUARE CANTILEVER PLATE

Young	Warburton	Finite Element	Finite Difference
 3.494	3.518	3.47	3.27
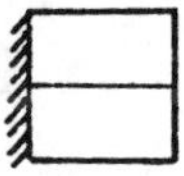 8.547	9.516	8.53	8.91
 21.44	22.029	21.67	18.11
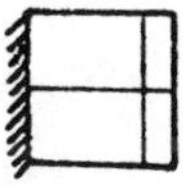 27.46	32.091	26.85	29.28
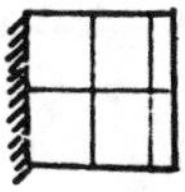 -	71.451	-	65.02

Figure 6.4

NON-DIMENSIONAL FREQUENCY FACTORS

RECTANGULAR CANTILEVER PLATE

	Warburton	Experimental	Finite Difference
	3.202	2.946	2.755
	16.487	15.180	-
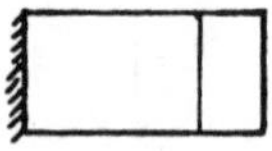	20.18	20.89	19.00
	46.43	47.38	36.53
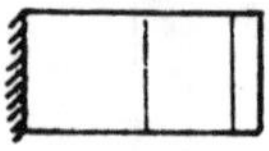	56.25	57.44	54.78
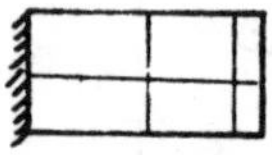	87.32	-	80.77
	87.20	91.24	76.17
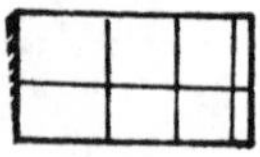	141.70	146.60	128.94

Figure 6.5

NON-DIMENSIONAL FREQUENCY FACTORS

RECTANGULAR PLATE - TWO OPPOSITE EDGES FREE -

ONE EDGE SIMPLY SUPPORTED - ONE EDGE ENCASTRE

Warburton	Experimental	Finite Difference
14.15	12.00	14.58
 29.14	22.72	29.58
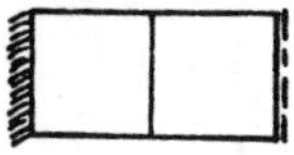 45.38	—	46.77
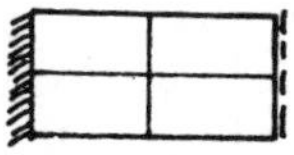 67.62	56.70	67.71
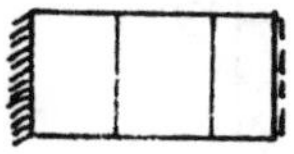 96.04	—	96.40
99.22	—	103.20
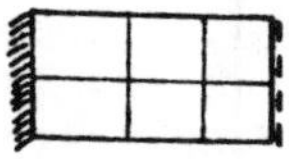 120.80	107.10	119.10
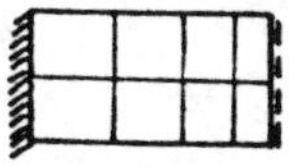 207.05	---	185.20

Figure 6.6

NON-DIMENSIONAL FREQUENCY FACTORS

2 OPPOSITE EDGES FREE

OTHERS SIMPLY SUPPORTED

Warburton	Experimental	Finite Difference
9.87	9.60	9.56
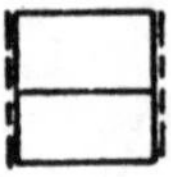 16.22	16.30	15.90
 36.82	34.00	36.22
 39.48	-	38.18
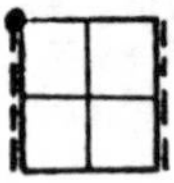 47.14	44.70	45.28
 71.33	68.99	68.95

Figure 6.7a

NON-DIMENSIONAL FREQUENCY FACTORS

2 OPPOSITE EDGES FREE

OTHERS SIMPLY SUPPORTED

Warburton	Experimental	Finite Difference
75.35	70.40	73.55
88.83	75.50	84.84
96.86	-	91.30
111.65	-	108.56
157.91	-	147.71
166.10	-	153.13

Figure 6.7b

NON-DIMENSIONAL FREQUENCY FACTORS

HOLE IN PLATE

2 OPPOSITE EDGES FREE

OTHERS SIMPLY SUPPORTED

Experimental	Finite Difference
7.722	7.538
14.247	16.524
35.998	38.52
36.798	39.488
79.827	79.110
-	80.843

Figure 6.8a

NON-DIMENSIONAL FREQUENCY FACTORS

HOLE IN PLATE

2 OPPOSITE EDGES FREE

OTHERS SIMPLY SUPPORTED

Experimental	Finite Difference
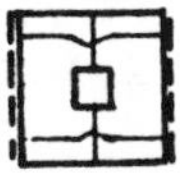 83.42	82.440
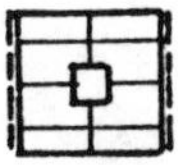 101.469	88.583
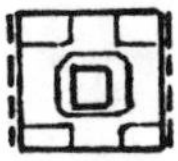 116.514	123.30
 128.441	134.055
 -	131.715
 -	135.38

Figure 6.8b

NON DIMENSIONAL FREQUENCY FACTORS

2 OPPOSITE EDGES FREE

OTHERS PARTIALLY SIMPLY SUPPORTED

Experimental.	Finite Difference
4.726	9.203
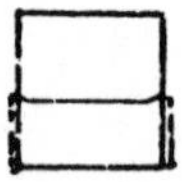 8.125	7.331
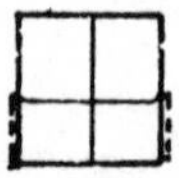 21.672	20.76
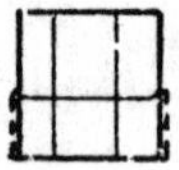 30.13	33.65
 37.44	35.59
 -	55.12
 -	66.6
 -	75.59

Figure 6.9

7. CONCLUSIONS

In the absence of an exact solution to the problem of plate vibration, an approximate method must be sought which can be applied as a general solution. Substantial advances have been made in this direction by the application of the Rayleigh-Ritz type of solution but, the limitations of this method prevent it from becoming a general solution. The alternative solution, by finite difference methods, appears capable of dealing with all of the types of plates soluble by the Rayleigh-Ritz method. The accuracy of both types of solution is comparable and it has been shown that the finite difference method is capable of handling a much wider range of plate configurations. This is substantiated in Section 6, by considering plates with central holes cut in them and plates having boundary conditions which vary along an edge of the plate.

The difference equations set up in Section 2, to correspond to the boundary and governing partial differential equations, can be refined to give a greater degree of accuracy. However, in this context, the boundary conditions must not be allowed to have a greater degree of truncation error than the governing differential equation or the solution may become singular.

From the computer results of Section 6, it can be seen that the finite difference equations and computer programmes constructed in this thesis, are capable of determining the coupled mode shapes for square encastre and free plates. The accuracy of the solution for the higher mode shapes is adequate provided at least three mesh lengths correspond to each nodal line.

Once the 'molecule' diagrams have been set up for the various types of boundary conditions, a computer can be used to build-up the molecule diagrams for any plate shape. The input to the computer need only be the size and shape of the plate, together with its physical properties. The computer programme would be able to select the optimum mesh size and then formulate and solve the problem incorporating any special techniques required for specific plate shapes.

From the experimental work of this paper, it can be seen that

From the experimental work of this paper, it can be seen that an encastre end to a plate could only be achieved if the thickness of the end support was at least five times the thickness of the plate and that the simply supported condition was not easy to achieve. It is therefore unlikely that the boundary condition of a plate in service would fall into one of the three characteristic boundary conditions but would be somewhere between these extremes. The finite difference method would be invaluable in such cases to analyse installations which have plate vibration problems. Finite difference equations can be built up super-imposing, in some pre-determined ratio, the separate effects of the three characteristic boundary conditions and obtain a solution for such a plate. This ratio can be varied until the resonant frequency and mode shape corresponds to that of the plate being investigated. A solution to the problem could then be obtained by applying the finite difference method to consider the effect of cutting holes in the plate or varying the boundary condition at specific locations. The position of the nodal lines could also be established by this method and used to mount items of equipment which must be vibration free.

The greatest source of uncertainty in the method is the corner points of free/free boundaries and the torsional modes of cantilevered plates. Some work in this field could produce a general solution to the problem of plate vibration.

8. REFERENCES

1. E.F.F. Chladni: Neue bietrage zur akustik, 1817.

2. Lord Rayleigh: Theory of sound, Vol. 1, 2nd Edition, Macmillan & Co., London, 1894.

3. W. Voight: Bemerkung zu dem problem der transversalen schwingungen rechteckiger platten. Nachrichten, Gottingen, 1893.

4. A.E.H. Love: Mathematical theory of elasticity. 4th Edition, Cambridge University Press, 1927.

5. G.B. Warburton: The vibration of rectangular plates. Proc. of the Inst. Mech. Eng., Vol. 168, 1954.

6. W.G. Bickley: Finite difference formulae for the square lattice. Quart. J. Mech. Appl. Maths., Vol. 1, 1948.

7. L. Collatz: Numerical treatment of differential equations. Springer-Verlag, Berlin (translation 1959).

8. S.C. Mathur: Numerical studies in partial differential equations of elliptic type. PhD Thesis, University of Glasgow, 1965.

9. W.E. Milne: Numerical solution of differential equations. John Wiley, London, 1957.

10. A.E. Heiba: Vibration characteristics of a cantilever plate with swept back leading edges. Coll. Aero. Cranfield Rep. No. 82, 1954.

11. M.V. Barton: Vibration of rectangular and skew cantilever plates. J. of Appl. Mech., Trans. A.S.M.E. Vol. 18, 1951.

Dr. R.F. McLean
Department of Dynamics and Control
University of Strathclyde
Glasgow, Scotland

Some problems in acoustic scattering

R. PETHRICK

ABSTRACT

Analysis of the swept frequency acoustic resonator and the problems associated with a more detailed modelling of the system are summarised. The use of ultrasonic waves in the exploration of scattering phenomena in heterogeneous polymer systems are also discussed.

INTRODUCTION

Approximately ten years ago the concept of a swept frequency acoustic resonator was introduced by Eggers [1]. His experiments and simple theory predicted the widths of the observed resonance peaks and the position of the major resonance features but were unable to explain the detail of the response function. In an attempt to model more accurately the behaviour of the acoustic resonator Dev et al [2] attempted to compute the response of a cavity constructed from two piezo electric crystals and an isotropic coupling media. A two dimensional analysis of the system indicated that in addition to the main resonance features additional resonances corresponding to overtones of the fundamental may be expected to be observed. A careful examination of the experimental response for this system confirms this prediction. At low frequency these overtones have little amplitude, but at higher frequency the width of the peak is sufficient to allow coupling with the first overtone. The overtone has the form of a radial mode, which is in contrast to the simple piston-like displacement associated with the fundamental. Experimentally the width of the peak is a combination of various effects. At low frequency, the finite size of the transducer leads to diffraction of the sound beam. At higher frequencies this loss becomes negligible and the width is determined purely by the damping factor of the liquid coupling media. In the region of the fundamental of the quartz crystal additional broadening of the resonance occurs as a consequence of piezo electric damping. At even higher frequencies the width is increased over that of the liquid damping due to coupling of the radial with the pure modes of propagation in the cavity. In addition, a broadening of resonances in the region of 3 MHz for a 5 MHz X-cut fundamental crystal can be associated with strong coupling of the radial mode with the AT fundamental of vibration. The simple models do not explain the details of the spectrum described above. It is important that a more detailed model be developed which explains the behaviour of cavity and also allows exploration of the analogous system in which the coupling media is an anisotropic solid. A simple analytical formula would allow a more detailed exploration of molecular relaxation process.

During the last five years studies at Strathclyde have been concerned with the investigation of polymers in which are suspended glass beads [3], polymers which contain air voids [4] and spherulite structures [5]. Using simple models it is possible to describe the gross features which are observed. It is, however, clear that these simple models do not explain the data obtained. One of the problems which is common to the scattering theories is that sharp boundaries are usually assumed. In a polymer system, surface adsorption leads to a high modulus even close to the glass surface and this is graded into the bulk. Similarly, scattering from spherulites involves a distribution of sizes and a media which is itself anisotropic. Thermal diffusion between spherulites must take a finite time and hence can lead to frequency dependent relaxation phenomena. It is clear that detailed topographical modelling of these systems will allow important molecular information to be derived from these studies.

REFERENCES

1. F. Eggers: Acustica 19, 323, 1968.
2. S.B. Dev, S. Sarker and R.A. Pethrick: J. Physics E. 6, 139, 1973.
3. P.K. Datta and R.A. Pethrick: J. Phys. D Appl. Phys. 13, 153, 1980.
4. K. Adachi, G. Harrison, J. Lamb, A.M. North and R.A. Pethrick: Polymer 22, 1026, 1981.
5. K. Adachi, G. Harrison, J. Lamb, A.M. North and R.A. Pethrick: Polymer 22, 1048, 1981.

Dr R.A. Pethrick
Department of Pure and Applied Chemistry
University of Strathclyde
Glasgow, Scotland.

Wave propagation in elastic rods; approximate solutions

G. EASON

SUMMARY

Approximate solutions of the linear equations of motion for an elastic rod proposed by Green and Laws are obtained. The special cases of a curved rod and a spiral spring are considered and it is shown that the approximate theory gives a good approximation to the more general equations.

1. INTRODUCTION

The propagation of waves in elastic rods has been studied for many years. The early approaches were usually based on assumptions which were not always well defined. In recent years there have been several derivations of the basic equations from the three dimensional theory of elasticity and also using director theory. These two approaches lead to the same equations, see for example Green et al. [1,2]. The equations adopted here are those due initially to Green and Laws [3] and further developed by Green et al. [4]. For simplicity it is assumed that isothermal conditions exist and that there are no body forces present.

The basic equations given in [3] are more general than the traditional equations to be found in Love [5], for example. In the present paper the connection between the two sets of equations is explored for infinitesimal deformations. It is shown that by making appropriate assumptions the classical equations can be obtained from those to be found in [3]. There are then additional equations which predict waves not given by the classical theory and which give rise to secondary effects. These secondary effects are expected to be most marked in rods with non-symmetric cross sections or with large initial curvatures. Modifications to the classical results may then be obtained. Two particular examples are considered as special cases of the general results. These are the propagation of waves in a curved rod and also in a spiral spring. Good agreement with previous results is obtained.

The basic equations due to Green and Laws [3] and devloped by Green at al. [4] are presented in section 2. In section 3 the general approximate equations for the linear theory are derived. The special cases of a curved rod, which has previously been considered by Eason [6] and others, is examined in section 4. Finally the application of this approach to a spiral spring is considered in section 5.

2. THE BASIC EQUATIONS

In this section the basic equations of the non-linear theory of elastic rods of Green and Laws [3] are presented and the corresponding linear equations are then derived. These basic equations may also be obtained from the three-dimensional equations of elasticity, see Green et al. [1].

A rod is defined to be a curve $\mathcal{c}$ embedded in Euclidean 3-space with equation

$$\underline{r} = \underline{r}(\theta,t), \tag{2.1}$$

where $\underline{r}$ is the position vector of a point on $\mathcal{c}$, t is the time and θ is a convected coordinate defining points on $\mathcal{c}$. At time t=0 the initial undeformed position of the rod, denoted by $\mathcal{C}$, has equation $\underline{R}=\underline{r}(\theta,0)$. In the following only small displacements from the initial curve $\mathcal{C}$ are considered.

Two assigned directors $\underline{a}_\alpha$ are assumed to exist at each point of $\mathcal{c}$ and their duals at t=0 are denoted by $\underline{A}_\alpha$. Here Greek indices take the values 1,2 and Latin indices the values 1,2,3. The orientation and magnitude of $\underline{A}_\alpha$ are not restricted at this stage but the subsequent analysis is greatly simplified if they are chosen to be unit vectors in the plane of a cross-section of the rod. A third vector, $\underline{A}_3$ at t=0 and $\underline{a}_3$ at time t is defined by

$$\underline{a}_3 = \frac{\partial \underline{r}}{\partial \theta}. \tag{2.2}$$

The $\underline{a}_i$ are assumed to depend on θ and t alone with $[\underline{a}_1,\underline{a}_2,\underline{a}_3]>0$.

A set of reciprocal base vectors $\underline{a}^i$ also exist with

$$\underline{a}^i . \underline{a}_j = \delta^i_j, \tag{2.3}$$

where δ^i_j is the Kronecker delta.

The quantities a_{ij}, a^{ij}, κ_{ij}, $\kappa_i^{.j}$ are introduced where, at time t,

$$a_{ij} = \underline{a}_i . \underline{a}_j, \tag{2.4}$$

$$a^{ij} a_{kj} = \delta^i_k, \tag{2.5}$$

$$\kappa_{ij} = \underline{a}_j . \frac{\partial \underline{a}_i}{\partial \theta}, \tag{2.6}$$

$$\kappa_i^{.j} = a^{rj} \kappa_{ir}. \tag{2.7}$$

At t=0 these quantities are denoted by A_{ij}, A^{ij}, K_{ij}, $K_i^{.j}$ respectively.

The equation of mass conservation is

$$\rho \sqrt{a_{33}} = \beta, \tag{2.8}$$

where ρ is the mass per unit length.

The equations of motion are

$$\frac{\partial n^i}{\partial \theta} + \kappa_r^{.i} n^r = \beta c^i, \tag{2.9}$$

$$\pi^{12} - \pi^{21} + p^{\gamma 2} \kappa_\gamma^{.1} - p^{\gamma 1} \kappa_\gamma^{.2} = 0, \tag{2.10}$$

$$\pi^{\beta 3} + p^{\alpha 3} \kappa_\alpha^{.\beta} - p^{\alpha\beta} \kappa_\alpha^{.3} - n^\beta = 0, \tag{2.11}$$

$$\pi^{\alpha i} = \beta q^{\alpha i} + \frac{\partial p^{\alpha i}}{\partial \theta} + \kappa_r^{.i} p^{\alpha r}, \tag{2.12}$$

where n^i are the force components, $p^{\alpha i}$ are the director force components, c^i are the components of acceleration and $q^{\alpha i}$ are the director inertia terms. In (2.9) and (2.12) the body force and assigned director force terms have been omitted for simplicity. The definition of $\pi^{\alpha i}$ used here differs from that used in [3].

Isothermal motions only are considered here so that, for an elastic rod, the Helmholtz free energy per unit mass, A, is given by

$$A = A(\gamma_{ij}, \sigma_{\alpha i}, A_{ij}, K_{\alpha i}), \tag{2.13}$$

where

$$\gamma_{ij} = a_{ij} - A_{ij}, \tag{2.14}$$

$$\sigma_{ij} = \kappa_{ij} - K_{ij}. \tag{2.15}$$

Once A is prescribed the stresses are given in terms of the strain and curvature components by

$$n^3 - p^{\alpha 3}\kappa_{\alpha}^{\cdot 3} = 2\beta \frac{\partial A}{\partial \gamma_{33}}, \tag{2.16}$$

$$n^\beta - p^{\alpha 3}\kappa_{\alpha}^{\cdot \beta} = \beta \frac{\partial A}{\partial \gamma_{\beta 3}}, \tag{2.17}$$

$$\pi^{\alpha\beta} + \pi^{\beta\alpha} - (p^{\gamma\beta}\kappa_{\gamma}^{\cdot \alpha} + p^{\gamma\alpha}\kappa_{\gamma}^{\cdot \beta}) = 4\beta \frac{\partial A}{\partial \gamma_{\alpha\beta}}, \tag{2.18}$$

$$p^{\alpha i} = \beta \frac{\partial A}{\partial \sigma_{\alpha i}}. \tag{2.19}$$

Equations (2.1)-(2.19) are the basic equations of the general non-linear theory for an isothermal elastic rod in the absence of body forces. These equations are now specialized to a linear theory in which all displacements from the initial configuration $\mathcal{C}$ are small.

For simplicity it is assumed that the $\underline{A}_i$ are orthogonal unit vectors so that

$$\underline{A}^i = \underline{A}_i, \tag{2.20}$$

$$A_{ij} = \delta_{ij}, \tag{2.21}$$

$$A^{ij} = \delta^{ij}, \tag{2.22}$$

$$K_i^{\cdot j} = K_{ij} = \underline{A}_j \cdot \frac{\partial \underline{A}_i}{\partial \theta}, \tag{2.23}$$

$$\underline{R} = \int_0^\theta \underline{A}_3 d\theta, \tag{2.24}$$

and it is assumed that one end of the rod coincides with $\theta=0$.

Green et al. [4] have considered an elastic rod which is subjected to an initial finite deformation followed by a small infinitesimal deformation. In the notation of [4], neglecting the initial deformation and defining ε to be a small real parameter it is found that

$$\underline{r} = \underline{R} + \varepsilon\underline{u} = \underline{R} + \varepsilon u_i \underline{A}_i, \tag{2.25}$$

$$\underline{a}_i = \underline{A}_i + \varepsilon\underline{b}_i = \underline{A}_i + \varepsilon b_{ij}\underline{A}_j, \tag{2.26}$$

$$\underline{a}^i = \underline{A}_i - \varepsilon b_j^{\cdot i}\underline{A}_j, \tag{2.27}$$

$$b_j^{\cdot i} = b_{ji}, \tag{2.28}$$

$$\underline{b}_3 = \frac{\partial \underline{u}}{\partial \theta}, \tag{2.29}$$

$$a_{ij} = A_{ij} + \varepsilon(b_{ij} + b_{ji}), \tag{2.30}$$

$$\gamma_{ij} = \varepsilon(b_{ij} + b_{ji}), \tag{2.31}$$

$$\kappa_{ij} = K_{ij} + \varepsilon\lambda_{ij}, \tag{2.32}$$

$$\sigma_{ij} = \varepsilon\lambda_{ij}, \tag{2.33}$$

$$\lambda_{ij} = \frac{\partial b_{ij}}{\partial \theta} + b_{ik}K_{kj} + b_{jk}K_{ik}, \tag{2.34}$$

$$\kappa_i^{\cdot j} = K_i^{\cdot j} + \varepsilon\mu_i^{\cdot j}, \tag{2.35}$$

$$\mu_i^{\cdot j} = \frac{\partial b_{ij}}{\partial \theta} + b_{ik}K_k^{\cdot j} - b_{kj}K_i^{\cdot k}, \tag{2.36}$$

where powers of ε above the first have been neglected and indices are written in the covariant position whenever possible.

All forces are assumed to be of order ε so that

$$n^i = \varepsilon\nu_i, \tag{2.37}$$

$$p^{\alpha i} = \varepsilon\xi_{\alpha i}, \tag{2.38}$$

$$\pi^{\alpha i} = \varepsilon\omega_{\alpha i}. \tag{2.39}$$

The equations of motion (2.9)-(2.12) now give

$$\frac{\partial \nu_1}{\partial \theta} + K_{11}\nu_1 + K_{21}\nu_2 + K_{31}\nu_3 = \beta \frac{\partial^2 u_1}{\partial t^2}, \tag{2.40}$$

$$\frac{\partial \nu_2}{\partial \theta} + K_{12}\nu_1 + K_{22}\nu_2 + K_{32}\nu_3 = \beta \frac{\partial^2 u_2}{\partial t^2}, \tag{2.41}$$

$$\frac{\partial \nu_3}{\partial \theta} + K_{13}\nu_1 + K_{23}\nu_2 + K_{33}\nu_3 = \beta \frac{\partial^2 u_3}{\partial t^2}, \tag{2.42}$$

$$\frac{\partial P}{\partial \theta} + (K_{11}+K_{22})P + K_{32}\xi_{13} - K_{31}\xi_{23} = \alpha_1 \frac{\partial^2 b_{12}}{\partial t^2} - \alpha_2 \frac{\partial^2 b_{21}}{\partial t^2}, \tag{2.43}$$

$$\frac{\partial \xi_{13}}{\partial \theta} - \nu_1 + K_{23}P + \xi_{13}(K_{33}+K_{11}) + K_{21}\xi_{23} = \alpha_1 \frac{\partial^2 b_{13}}{\partial t^2}, \tag{2.44}$$

$$\frac{\partial \xi_{23}}{\partial \theta} - \nu_2 - K_{13}P + K_{12}\xi_{13} + \xi_{23}(K_{22}+K_{33}) = \alpha_2 \frac{\partial^2 b_{23}}{\partial t^2}, \tag{2.45}$$

$$\omega_{11} = \frac{\partial \xi_{11}}{\partial \theta} + K_{11}\xi_{11} + \tfrac{1}{2}K_{21}(P+Q) + K_{31}\xi_{13} - \alpha_1 \frac{\partial^2 b_{11}}{\partial t^2}, \tag{2.46}$$

$$\omega_{12} + \omega_{21} = \frac{\partial Q}{\partial \theta} + K_{12}\xi_{11} + \tfrac{1}{2}K_{22}(P+Q) + K_{32}\xi_{13}$$

$$+ \tfrac{1}{2}K_{11}(Q-P) + K_{21}\xi_{22} + K_{31}\xi_{23} - \alpha_1 \frac{\partial^2 b_{12}}{\partial t^2} - \alpha_2 \frac{\partial^2 b_{21}}{\partial t^2}, \tag{2.47}$$

$$\omega_{22} = \frac{\partial \xi_{22}}{\partial \theta} + \tfrac{1}{2}K_{12}(Q-P) + K_{22}\xi_{22} + K_{32}\xi_{23} - \alpha_2 \frac{\partial^2 b_{22}}{\partial t^2}, \tag{2.48}$$

where

$$P = \xi_{12} - \xi_{21}, \tag{2.49}$$

$$Q = \xi_{12} + \xi_{21}. \tag{2.50}$$

The right hand sides of (2.40)-(2.42) are the traditional inertia terms and the usual assumptions have been made concerning the director inertia terms in (2.43)-(2.48).

A brief examination of (2.40)-(2.48) indicates that (2.40)-(2.45) are the classical equations of motion for an elastic rod to be found for example in Love [5]. Equations (2.46)-(2.48) are additional equations which relate to the motion of the cross-section of the rod. Since the classical equations give a satisfactory description of the motion in most cases it is to be expected that ξ_{11}, ξ_{22} and Q will be smaller in magnitude than ξ_{13} and ξ_{23}. It is this conjecture that will be explored here.

In order to determine specific constitutive equations from (2.16)-(2.19) it is necessary to prescribe the free energy A. The form adopted here is that used by Green et al. [4] and Eason [6] in which A is assumed to depend on γ_{ij} and $\sigma_{\alpha i}$ alone. The most general quadratic form for A (to give a linear theory) is then

$$2\beta A = k_1\gamma_{11}^2 + k_2\gamma_{22}^2 + k_3\gamma_{33}^2 + \tfrac{1}{4}k_4(\gamma_{12}+\gamma_{21})^2$$

$$+ k_5\gamma_{23}^2 + k_6\gamma_{13}^2 + k_7\gamma_{11}\gamma_{22} + k_8\gamma_{11}\gamma_{33} + k_9\gamma_{22}\gamma_{33}$$

$$+ k_{10}\sigma_{11}^2 + k_{11}\sigma_{22}^2 + k_{12}\sigma_{12}^2 + k_{13}\sigma_{21}^2 + k_{14}\sigma_{12}\sigma_{21}$$

$$+ k_{15}\sigma_{23}^2 + k_{16}\sigma_{13}^2 + k_{17}\sigma_{11}\sigma_{22}, \tag{2.51}$$

where k_1-k_{17} are assumed to be constant. Equations (2.16)-(2.19) now give

$$\nu_3 - \xi_{13}K_{13} - \xi_{23}K_{23} = 2k_8 b_{11} + 2k_9 b_{22} + 4k_3 b_{33}, \tag{2.52}$$

$$\nu_1 - \xi_{13}K_{11} - \xi_{23}K_{21} = k_6(b_{13}+b_{31}), \tag{2.53}$$

$$\nu_2 - \xi_{13}K_{12} - \xi_{23}K_{22} = k_5(b_{23}+b_{32}), \tag{2.54}$$

$$\omega_{11} - \xi_{11}K_{11} - \xi_{21}K_{21} = 4k_1 b_{11} + 2k_7 b_{22} + 2k_8 b_{33}, \tag{2.55}$$

$$\omega_{22} - \xi_{12}K_{12} - \xi_{22}K_{22} = 2k_7 b_{11} + 4k_2 b_{22} + 2k_9 b_{33}, \tag{2.56}$$

$$\omega_{12} + \omega_{21} - \xi_{12}K_{11} - \xi_{22}K_{21} - \xi_{11}K_{12} - \xi_{21}K_{22} = 2k_4(b_{12}+b_{21}), \tag{2.57}$$

$$\xi_{11} = k_{10}\lambda_{11} + \tfrac{1}{2}k_{17}\lambda_{22}, \tag{2.58}$$

$$P = k_{12}\lambda_{12} - k_{13}\lambda_{21} - \tfrac{1}{2}k_{14}(\lambda_{12}-\lambda_{21}), \tag{2.59}$$

$$\xi_{13} = k_{16}\lambda_{13}, \tag{2.60}$$

$$Q = k_{12}\lambda_{12} + k_{13}\lambda_{21} + \tfrac{1}{2}k_{14}(\lambda_{12}+\lambda_{21}), \tag{2.61}$$

$$\xi_{22} = k_{11}\lambda_{22} + \tfrac{1}{2}k_{17}\lambda_{11}, \tag{2.62}$$

$$\xi_{23} = k_{15}\lambda_{23}. \tag{2.63}$$

As was the case with the equations of motion, certain of thse equations are analogous to equations to be found in the classical theory of Love [5]. The consequences of neglecting ξ_{11}, ξ_{22} and Q, and their associated variables, as a first approximation to these general equations will be explored in the next section.

3. THE APPROXIMATE THEORY

It has already been indicated in the previous section that the basic equations fall into two groups. The first group consisting of the twelve equations (2.40)-(2.45), (2.52)-(2.54), (2.59)-(2.60) and (2.63) corresponds to the classical equations for an elastic rod, see for example Love [5]. The second group of nine equations (2.46)-(2.48), (2.55)-(2.58), (2.61) and (2.62) is new and represents motion associated with deformation of the cross section.

The two groups of equations are not completely independent. The variables of interest in the main group are ν_1, ν_2, ν_3, P, ξ_{13}, ξ_{23}, b_{13}, b_{23}, $B_1 = b_{12} - b_{21}$, u_1, u_2 and u_3. The second group consists of ω_{11}, ω_{22}, $\omega_{12} + \omega_{21}$, ξ_{11}, ξ_{22}, Q, b_{11}, b_{22} and $B_2 = b_{12} + b_{21}$. Equations (2.43), (2.52), (2.59), (2.60) and (2.61) of the main group contain variables from the second group and all equations of the second group contain variables from the first group. The linking of the two groups of equations suggests that in principle the two motions will be linked although in most particular situations some of the linking terms will be zero owing to certain initial curvatures being zero or symmetries being present.

When the linking terms are neglected the first group of equations consists of (2.40)-(2.42), (2.44), (2.45), (2.53), (2.54) together with

$$\frac{\partial P}{\partial \theta} + (K_{11}+K_{22})P + K_{32}\xi_{13} - K_{31}\xi_{23} = \gamma \frac{\partial^2 B_1}{\partial t^2}, \tag{3.1}$$

$$\nu_3 - \xi_{13}K_{13} - \xi_{23}K_{23} = 2k_3 b_{33}, \tag{3.2}$$

$$\begin{aligned} P = {} & \tfrac{1}{2}(k_{12}+k_{13}-k_{14}) \frac{\partial B_1}{\partial \theta} + \tfrac{1}{2}B_1(k_{12}-k_{13})(K_{22}-K_{11}) \\ & + b_{13}\{(k_{12}-\tfrac{1}{2}k_{14})K_{32} - (k_{13}-\tfrac{1}{2}k_{14})K_{23}\} \\ & + b_{23}\{(k_{12}-\tfrac{1}{2}k_{14})K_{13} - (k_{13}-\tfrac{1}{2}k_{14})K_{31}\}, \end{aligned} \tag{3.3}$$

$$\xi_{13} = k_{16}\left(\frac{\partial b_{13}}{\partial \theta} + \tfrac{1}{2}B_1K_{23} + b_{13}K_{33} + b_{31}K_{11} + b_{32}K_{12} + b_{33}K_{13}\right), \tag{3.4}$$

$$\xi_{23} = k_{15}\left(\frac{\partial b_{23}}{\partial \theta} - \tfrac{1}{2}B_1K_{13} + b_{23}K_{33} + b_{31}K_{21} + b_{32}K_{22} + b_{33}K_{23}\right), \tag{3.5}$$

where

$$\gamma = \tfrac{1}{2}(\alpha_1 + \alpha_2). \tag{3.6}$$

The second group of equations is found to be

$$\frac{\partial \xi_{11}}{\partial \theta} - 4k_1 b_{11} - 2k_7 b_{22} = \alpha_1 \frac{\partial^2 b_{11}}{\partial t^2}, \tag{3.7}$$

$$\frac{\partial \xi_{22}}{\partial \theta} - 2k_7 b_{11} - 4k_2 b_{22} = \alpha_2 \frac{\partial^2 b_{22}}{\partial t^2}, \tag{3.8}$$

$$\frac{\partial Q}{\partial \theta} - 2k_4 B_2 = \gamma \frac{\partial^2 B_2}{\partial t^2}, \tag{3.9}$$

$$\begin{aligned} \xi_{11} = {} & k_{10}\left(\frac{\partial b_{11}}{\partial \theta} + 2b_{11}K_{11}\right) + \tfrac{1}{2}k_{17}\left(\frac{\partial b_{22}}{\partial \theta} + 2b_{22}K_{22}\right) \\ & + \tfrac{1}{2}(k_{10}+\tfrac{1}{2}k_{17})B_2(K_{12}+K_{21}), \end{aligned} \tag{3.10}$$

$$\begin{aligned} \xi_{22} = {} & \tfrac{1}{2}k_{17}\left(\frac{\partial b_{11}}{\partial \theta} + 2b_{11}K_{11}\right) + k_{11}\left(\frac{\partial b_{22}}{\partial \theta} + 2b_{22}K_{22}\right) \\ & + \tfrac{1}{2}(k_{11}+\tfrac{1}{2}k_{17})B_2(K_{12}+K_{21}), \end{aligned} \tag{3.11}$$

$$Q = \tfrac{1}{2}(k_{12}+k_{13}+k_{14})\frac{\partial B_2}{\partial\theta} + b_{11}\{k_{12}K_{12}+k_{13}K_{21}+\tfrac{1}{2}k_{14}(K_{12}+K_{21})\}$$

$$+ b_{22}\{k_{12}K_{12}+k_{13}K_{21}+\tfrac{1}{2}k_{14}(K_{12}+K_{21})\}$$

$$+ \tfrac{1}{2}B_2(k_{12}+k_{13}+k_{14})(K_{11}+K_{22}). \qquad (3.12)$$

In obtaining (3.1)-(3.12) it has been assumed that α_1 is close to α_2, k_{12} to k_{13} etc. and that several of the curvatures are small. Once solutions of each of these two groups of equations have been obtained it is possible to return to the full equations, obtain a second order set of equations similar to the first group and then use these to obtain modified values for the frequency equations. In rods with a significant amount of initial curvature or with significantly non-symmetrical cross-sections the approximations used here (and the classical equations) cannot be expected to lead to completely accurate results. They should, however, give reasonable first approximations to the correct values in most cases.

Some special cases will now be considered.

4. CURVED ROD

As a particular example of the application of the equations derived in sections 2 and 3 consider a rod whose centre-line forms the arc of a circle. This problem has been considered in some detail by Eason [6] who cites other references. Let a = 1/c denote the radius of the rod which lies initially in the plane defined by the unit vectors $\underline{A}_1$ and $\underline{A}_3$. Then

$$K_{31} = -K_{13} = K_{3}^{\cdot 1} = -K_{1}^{\cdot 3} = c, \qquad (4.1)$$

and all other K_{ij} and $K_{j}^{\cdot i}$ are zero.

The main group of equations results in

$$\frac{\partial \nu_1}{\partial\theta} + c\nu_3 = \beta\,\frac{\partial^2 u_1}{\partial t^2}, \qquad (4.2)$$

$$\frac{\partial \nu_2}{\partial\theta} = \beta\,\frac{\partial^2 u_2}{\partial t^2}, \qquad (4.3)$$

$$\frac{\partial \nu_3}{\partial\theta} - c\nu_1 = \beta\,\frac{\partial^2 u_3}{\partial t^2}, \qquad (4.4)$$

$$\frac{\partial P}{\partial \theta} - c\xi_{23} = \gamma \frac{\partial^2 B_1}{\partial t^2}, \tag{4.5}$$

$$\frac{\partial \xi_{13}}{\partial \theta} - \nu_1 = \alpha_1 \frac{\partial^2 b_{13}}{\partial t^2}, \tag{4.6}$$

$$\frac{\partial \xi_{23}}{\partial \theta} - \nu_2 + cP = \alpha_2 \frac{\partial^2 b_{23}}{\partial t^2}, \tag{4.7}$$

$$\nu_1 = k_6(b_{13}+b_{31}), \tag{4.8}$$

$$\nu_2 = k_5(b_{23}+b_{32}), \tag{4.9}$$

$$\nu_3 = 4k_3 b_{33} - c\xi_{13}, \tag{4.10}$$

$$P = (k_{12}+k_{13}-k_{14})(\tfrac{1}{2} \frac{\partial B_1}{\partial \theta} - cb_{23}), \tag{4.11}$$

$$\xi_{13} = k_{16}(\frac{\partial b_{13}}{\partial \theta} - cb_{33}), \tag{4.12}$$

$$\xi_{23} = k_{15}(\frac{\partial b_{23}}{\partial \theta} + \tfrac{1}{2}B_1 c). \tag{4.13}$$

These equations fall into a set governing extensional-flexural motion (4.2), (4.4), (4.6), (4.8), (4.10), (4.12) and a set governing torsional-flexural motion (4.3), (4.5), (4.7), (4.9), (4.11), (4.13).

When the constitutive equations are substituted into the equations of motion and solutions of the form

$$f(\theta,t) = \hat{f}(\xi,\omega)e^{i(\xi\theta-\omega t)}, \tag{4.14}$$

assumed it is found that for extensional-flexural motion

$$\begin{vmatrix} (k_6\xi^2+c^2e_1-\beta\omega^2) & \xi e_2 & \xi e_3 \\ \xi e_2 & (e_1\xi^2+c^2k_6-\beta\omega^2) & -c(k_{16}\xi^2-k_6) \\ \xi e_3 & -c(k_{16}\xi^2-k_6) & (k_{16}\xi^2+k_6-\alpha_1\omega^2) \end{vmatrix} = 0, \tag{4.15}$$

where

$$e_1 = 4k_3+c^2k_{16},\ e_2 = c(e_1+k_6),\ e_3 = k_6-c^2k_{16}. \tag{4.16}$$

Equation (4.15) corresponds to three curves in the ξ-ω plane, one of which passes through the origin and is tangent to the ξ-axis when ξ=c. The other two curves have cut-off frequencies

$$\omega_3 = \left[\frac{k_6}{\beta\alpha_1}(\beta+c^2\alpha_1)\right]^{\frac{1}{2}},\ \omega_4 = c(\frac{e_1}{\beta})^{\frac{1}{2}}. \tag{4.17}$$

The three curves are asymptotic to lines with slopes ζ_4,ζ_5,ζ_6 where

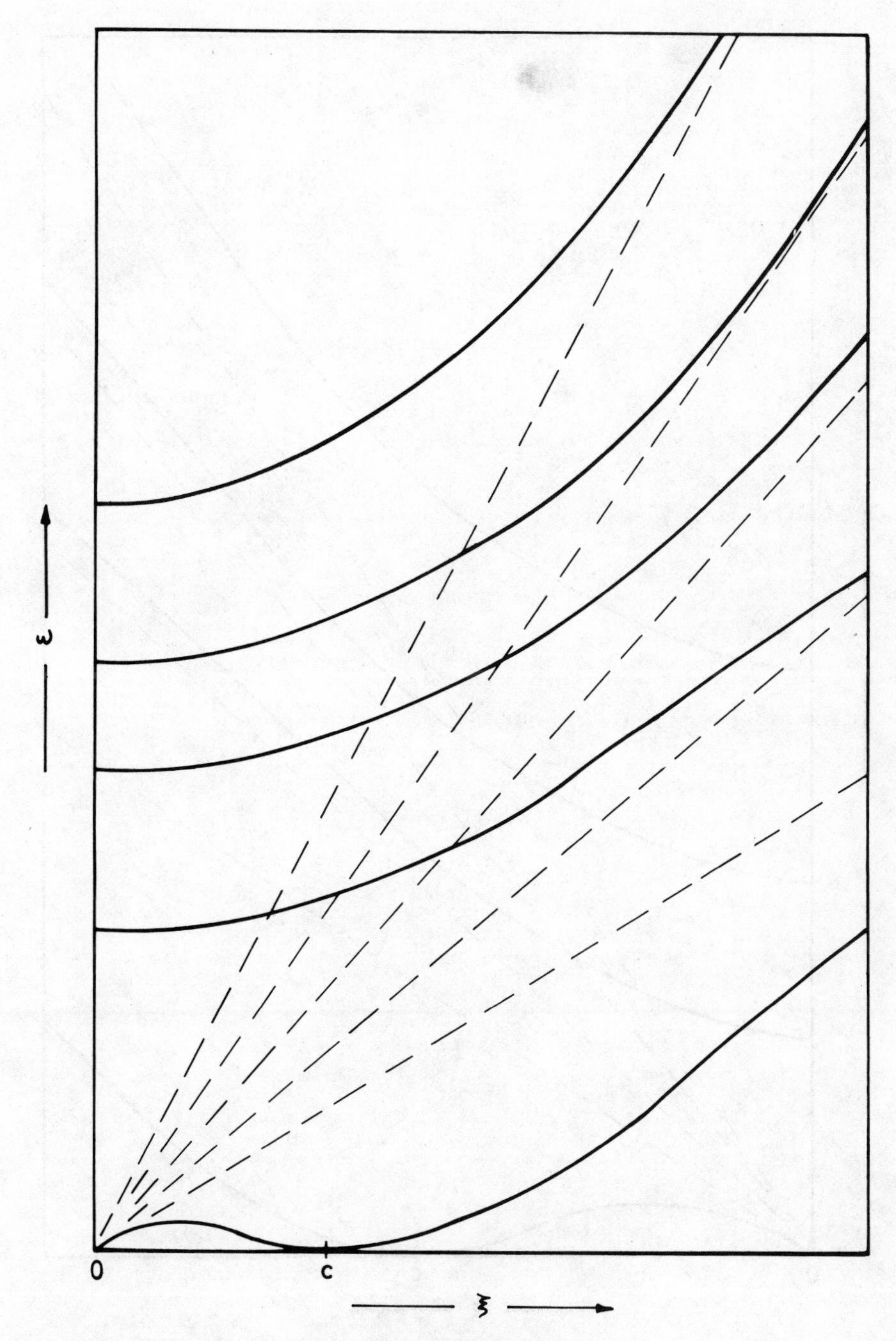

Figure 1. Variation of ω with ξ; extensional-flexural motion, curved rod.

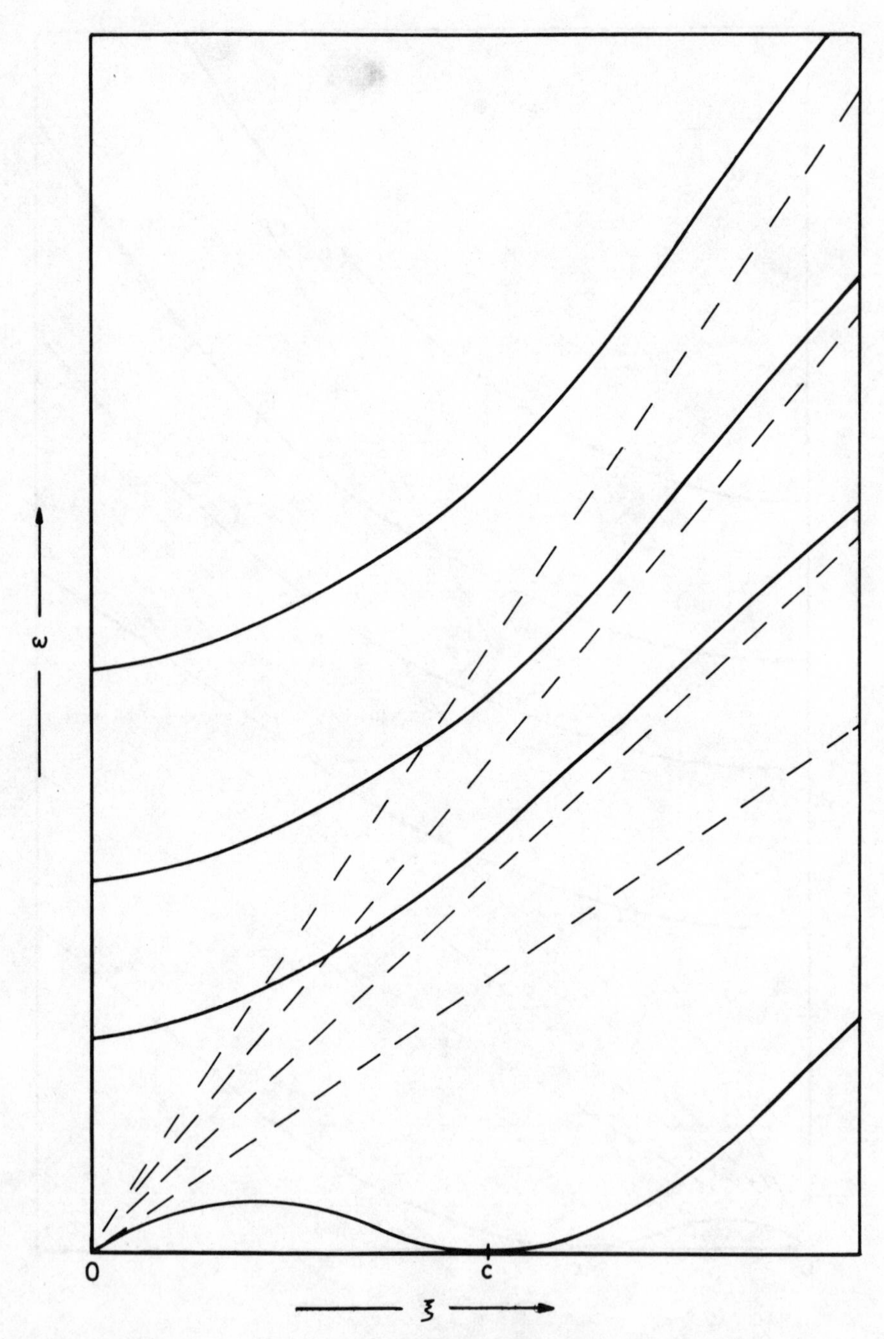

Figure 2. Variation of ω with ξ; torsional-flexural motion, curved rod.

$$\zeta_{4,5} = \left\langle \frac{1}{2\alpha_1\beta}\{(k_{16}\beta + e_1\alpha_1) \pm \left[(k_{16}\beta + e_1\alpha_1)^2 - 16\alpha_1 k_3 k_{16}\right]^{\frac{1}{2}}\}\right\rangle^{\frac{1}{2}}, \tag{4.18}$$

$$\zeta_6 = \left(\frac{k_6}{\beta}\right)^{\frac{1}{2}}, \tag{4.19}$$

and ζ_4 is taken to correspond to the positive sign in (4.18). The corresponding curves are given schematically by the first three curves in Figure 1.

The torsional-flexural equations, with (4.14), result in

$$\begin{vmatrix} (k_6\xi^2 - \rho\omega^2) & \xi k_5 & 0 \\ \xi k_5 & (k_{15}\xi^2 + d_1 - \alpha_2\omega^2) & \xi d_5 \\ 0 & \xi d_5 & (d_6\xi^2 + c^2 k_{15} - 2\gamma\omega^2) \end{vmatrix} = 0, \tag{4.20}$$

where

$$d_1 = k_5 + c^2 d_6,\ d_5 = c(k_{15} + d_6),\ d_6 = k_{12} + k_{13} - k_{14}. \tag{4.21}$$

Equation (4.20) gives three curves in the ξ-ω plane one of which passes through the origin and is tangent to the ξ-axis when $\xi = c$. The remaining two curves have cut-off frequencies given by

$$\omega_1 = \left(\frac{d_1}{\alpha_2}\right)^{\frac{1}{2}},\ \omega_2 = c\left(\frac{k_{15}}{2\gamma}\right)^{\frac{1}{2}}. \tag{4.22}$$

The three curves are asymptotic to lines with slopes $\zeta_1, \zeta_2, \zeta_3$ where

$$\zeta_1 = \left(\frac{k_5}{\beta}\right)^{\frac{1}{2}},\ \zeta_2 = \left(\frac{k_{15}}{\alpha_2}\right)^{\frac{1}{2}},\ \zeta_3 = \left(\frac{d_6}{2\gamma}\right)^{\frac{1}{2}}. \tag{4.23}$$

The corresponding curves are the first three curves in Figure 2.

The second group of equations (3.7)-(3.12) reduces to (3.7)-(3.9) together with

$$\xi_{11} = k_{10}\frac{\partial b_{11}}{\partial\theta} + \tfrac{1}{2}k_{17}\frac{\partial b_{22}}{\partial\theta}, \tag{4.24}$$

$$\xi_{22} = \tfrac{1}{2}k_{17}\frac{\partial b_{11}}{\partial\theta} + k_{11}\frac{\partial b_{22}}{\partial\theta}, \tag{4.25}$$

$$Q = \tfrac{1}{2}d_2\frac{\partial B_2}{\partial\theta}, \tag{4.26}$$

where

$$d_2 = k_{12} + k_{13} + k_{14}.$$

Proceeding in the same way as for the main group of equations it is found that additional extensional-flexural waves are given by

$$\begin{vmatrix} (k_{10}\xi^2+4k_1-\alpha_1\omega^2) & (\tfrac{1}{2}k_{17}\xi^2+2k_7) \\ (\tfrac{1}{2}k_{17}\xi^2+2k_7) & (k_{11}\xi^2+4k_2-\alpha_2\omega^2) \end{vmatrix} = 0. \qquad (4.27)$$

There are cut-off frequencies

$$\omega_{8,9} = \left\langle \frac{2}{\alpha_1\alpha_2}\{(\alpha_1 k_2+\alpha_2 k_1) \pm \left[(\alpha_1 k_2-\alpha_2 k_1)^2+\alpha_1\alpha_2 k_7^2\right]^{\frac{1}{2}}\}\right\rangle^{\frac{1}{2}}, \qquad (4.28)$$

where ω_8 is taken to have the positive sign and the two curves are asymptotic to lines with slopes

$$\zeta_{9,10} = \left\langle \frac{1}{2\alpha_1\alpha_2}\{(\alpha_1 k_{11}+\alpha_2 k_{10}) \pm \left[(\alpha_1 k_{11}-\alpha_2 k_{10})^2+k_{17}^2\right]^{\frac{1}{2}}\}\right\rangle^{\frac{1}{2}}. \qquad (4.29)$$

They are given by the top two curves in Figure 1. The torsional-flexural mode is given by

$$\omega = \{\frac{1}{2\gamma}\ (d_2\xi^2+4k_4)\}^{\frac{1}{2}}, \qquad (4.30)$$

with cut-off frequency

$$\omega_5 = (\frac{2k_4}{\gamma})^{\frac{1}{2}}. \qquad (4.31)$$

The curve is asymptotic to a line with slope

$$\zeta_7 = (\frac{d_2}{2\gamma})^{\frac{1}{2}}, \qquad (4.32)$$

and it is given by the top curve in Figure 2.

The results presented here give a good first approximation to the general results obtained in [6]. More refined results can be obtained by returning to the equations of section 2.

The particular case of the straight rod is obtained by taking $c=0$. The results may then be compared with those obtained by Green et al. [7].

5. THE SPIRAL SPRING

The second example considered is that of a spiral spring in the form of a helix of radius r and constant angle α to the OXY plane. The axis of the spring is chosen to lie along the z-axis so that at time $t=0$

$$\underline{R} = \underline{i}\ r\cos\psi + \underline{j}\ r\sin\psi + \underline{k}\ \theta\sin\alpha, \qquad (5.1)$$

where $\underline{i}$, $\underline{j}$, $\underline{k}$ are unit vectors in the directions OX, OY, OZ and

$$\psi = \theta\,\phi, \qquad (5.2)$$

$$\phi = (\cos\alpha)/r. \qquad (5.3)$$

The initial unit vectors are given by

$$\left.\begin{aligned}
\underline{A}_1 &= -\underline{i}\cos\psi - \underline{j}\sin\psi,\\
\underline{A}_2 &= \underline{i}\sin\alpha\sin\psi - \underline{j}\sin\alpha\cos\psi + \underline{k}\cos\alpha,\\
\underline{A}_3 &= -\underline{i}\cos\alpha\sin\psi + \underline{j}\cos\alpha\cos\psi + \underline{k}\sin\alpha,
\end{aligned}\right\} \tag{5.4}$$

so that

$$A_{ij} = \delta_{ij}, \tag{5.5}$$

$$K_{12} = -K_{21} = K_1^{\cdot 2} = -K_2^{\cdot 1} = \phi\sin\alpha = \eta_1, \tag{5.6}$$

$$K_{31} = -K_{13} = K_3^{\cdot 1} = -K_1^{\cdot 3} = \phi\cos\alpha = \eta_2, \tag{5.7}$$

and the remaining K_{ij} and $K_j^{\cdot i}$ are zero.

The main group of equations gives

$$\frac{\partial\nu_1}{\partial\theta} - \eta_1\nu_2 + \eta_2\nu_3 = \beta\,\frac{\partial^2 u_1}{\partial t^2}, \tag{5.8}$$

$$\frac{\partial\nu_2}{\partial\theta} + \eta_1\nu_1 = \beta\,\frac{\partial^2 u_2}{\partial t^2}, \tag{5.9}$$

$$\frac{\partial\nu_3}{\partial\theta} - \eta_2\nu_1 = \beta\,\frac{\partial^2 u_3}{\partial t^2}, \tag{5.10}$$

$$\frac{\partial P}{\partial\theta} - \eta_2\xi_{23} = \gamma\,\frac{\partial^2 B_1}{\partial t^2}, \tag{5.11}$$

$$\frac{\partial\xi_{13}}{\partial\theta} - \nu_1 - \eta_1\xi_{23} = \alpha_1\,\frac{\partial^2 b_{13}}{\partial t^2}, \tag{5.12}$$

$$\frac{\partial\xi_{23}}{\partial\theta} - \nu_2 + \eta_2 P + \eta_1\xi_{13} = \alpha_2\,\frac{\partial^2 b_{23}}{\partial t^2}, \tag{5.13}$$

$$\nu_1 = -\eta_1\xi_{23} + k_6(b_{13}+b_{31}), \tag{5.14}$$

$$\nu_2 = \eta_1\xi_{13} + k_5(b_{23}+b_{32}), \tag{5.15}$$

$$\nu_3 = -\eta_2\xi_{13} + 4k_3 b_{33}, \tag{5.16}$$

$$P = (k_{12}+k_{13}-k_{14})\left(\tfrac{1}{2}\,\frac{\partial B_1}{\partial\theta} - \eta_2 b_{23}\right), \tag{5.17}$$

$$\xi_{13} = k_{16}\left(\frac{\partial b_{13}}{\partial\theta} + \eta_1 b_{32} - \eta_2 b_{33}\right), \tag{5.18}$$

$$\xi_{23} = k_{15}\left(\frac{\partial b_{23}}{\partial\theta} + \tfrac{1}{2}\eta_2 B_1 - \eta_1 b_{31}\right). \tag{5.19}$$

In this case all twelve equations are linked together and there is no obvious separation as was the case for the curved rod.

By using the same methods as those adopted in section 4 it is found that

$$\begin{vmatrix} (G_1\xi^2+G_2-\beta\omega^2) & \xi\eta_1G_3 & \xi\eta_2G_4 & \eta_1\eta_2\xi k_{15} & -\xi G_5 & \eta_1(\xi^2k_{15}-k_5) \\ \xi\eta_1G_3 & (G_6\xi^2+\eta_1^2G_1-\beta\omega^2) & \eta_1\eta_2(k_{16}\xi^2+G_1) & \eta_1^2\eta_2k_{15} & \eta_1(k_{16}\xi^2-k_6) & -\xi G_7 \\ \xi\eta_2G_4 & \eta_1\eta_2(k_{16}\xi^2+G_1) & (G_8\xi^2+\eta_2^2G_1-\beta\omega^2) & \eta_1\eta_2^2k_{15} & \eta_2(k_{16}\xi^2-k_6) & \eta_1\eta_2\xi k_{15} \\ \xi\eta_1\eta_2k_{15} & \eta_1^2\eta_2k_{15} & \eta_1\eta_2^2k_{15} & (G_9\xi^2+\eta_2^2k_{15}-2\gamma\omega^2) & 0 & \xi\eta_2G_{10} \\ -\xi G_5 & \eta_1(k_{16}\xi^2-k_6) & \eta_2(k_{16}\xi^2-k_6) & 0 & (k_{16}\xi^2+k_6-\alpha_1\omega^2) & 0 \\ \eta_1(k_{15}\xi^2-k_5) & -\xi G_7 & \eta_1\eta_2\xi k_{15} & \xi\eta_2G_{10} & 0 & (k_{15}\xi^2+G_{11}-\alpha_2\omega^2) \end{vmatrix} = 0, \quad (5.20)$$

where

$$\begin{aligned} G_1 &= \eta_1^2k_{15} + k_6, & G_7 &= k_5 - \eta_1^2k_{15}, \\ G_2 &= \eta_1^2k_5 + 4\eta_2^2k_3 + \phi^4k_{16}, & G_8 &= 4k_3 + \eta_2^2k_{16}, \\ G_3 &= G_1 + k_5 + \phi^2k_{16}, & G_9 &= k_{12} + k_{13} - k_{14}, \\ G_4 &= G_1 + 4k_3 + \phi^2k_{16}, & G_{10} &= G_9 + k_{15}, \\ G_5 &= k_6 - \phi^2k_{16}, & G_{11} &= k_5 + \eta_2^2G_9. \\ G_6 &= k_5 + \eta_1^2k_{16}, & & \end{aligned} \quad (5.21)$$

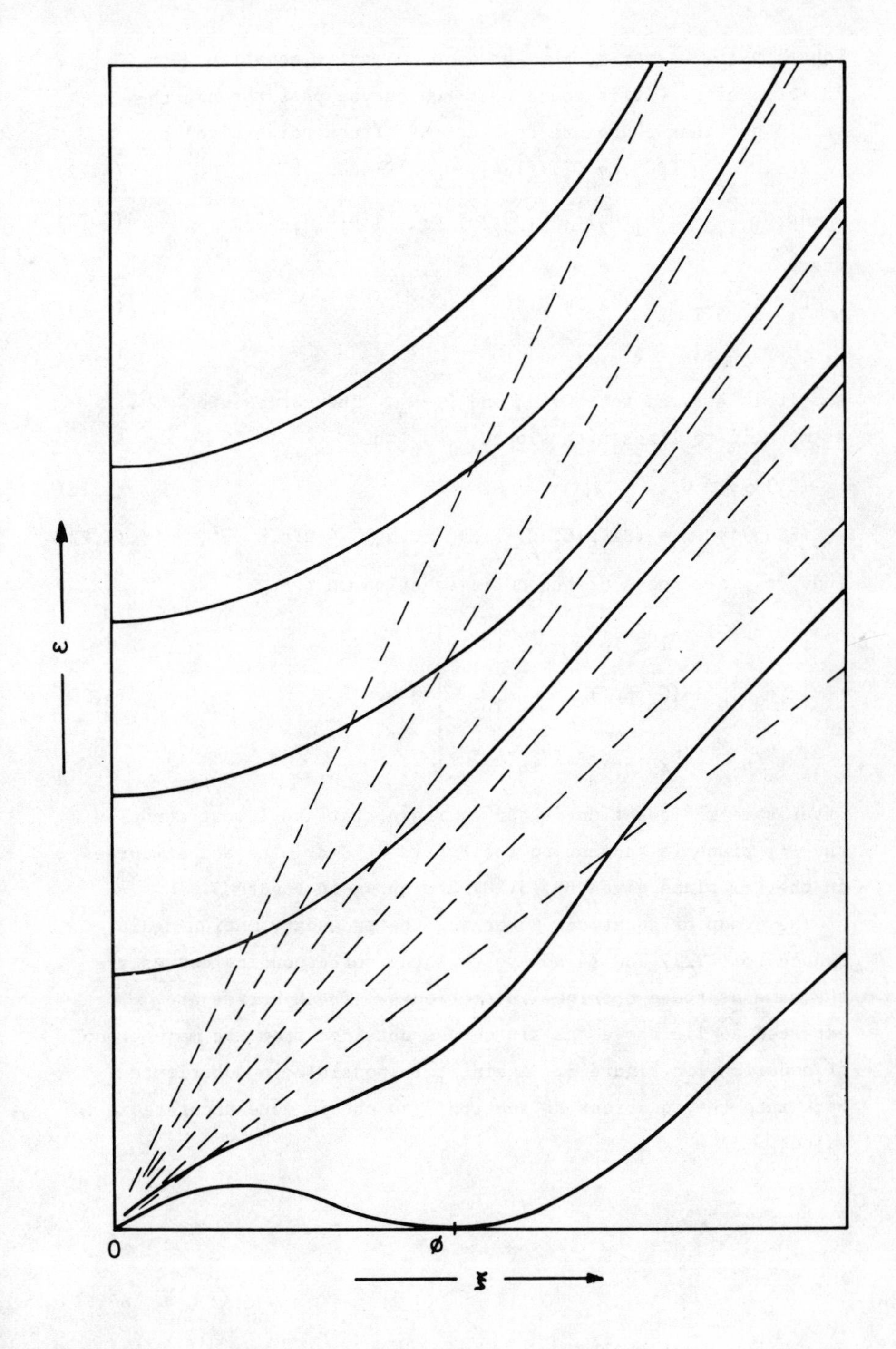

Figure 3. Variation of ω with ξ; spiral spring.

Equation (5.20) may be thought of as a sextic equation for ω^2 in terms of ξ. It is found that two curves pass through the origin and that there are four cut-off frequencies given by

$$2\beta\alpha_2\Omega_{1,2}^2 = (\beta G_{11}+\alpha_2 G_2)\pm\{(\beta G_{11}-\alpha_2 G_2)^2+4\beta\alpha_2\eta_1^2 k_5^2\}^{\frac{1}{2}}, \tag{5.22}$$

$$4\beta\gamma\alpha_1\Omega_{3,4}^2 = (H_1+H_2)\pm\{(H_1-H_2)^2+16\gamma^2\alpha_1^2\phi^4\eta_1^2 k_6 k_{15}\}^{\frac{1}{2}}, \tag{5.23}$$

where

$$H_1 = k_{15}\alpha_1(\eta_2^2\beta + 2\eta_1^2\phi^2\gamma), \tag{5.24}$$

$$H_2 = 2k_6\gamma(\beta + \phi^2\alpha_1), \tag{5.25}$$

and it is assumed that $\Omega_1>\Omega_2$ and $\Omega_3>\Omega_4$. The curves are asymptotic to lines with slopes $\chi_1-\chi_6$ where

$$(2\gamma)^{\frac{1}{2}}\chi_1 = G_9^{\frac{1}{2}}, \tag{5.26}$$

$$(2\beta\alpha_2)^{\frac{1}{2}}\chi_{2,3} = (\beta k_{15}+G_1\alpha_2)\pm\{(\beta k_{15}-G_1\alpha_2)^2+4\beta\alpha_2\eta_1^2 k_{15}^2\}^{\frac{1}{2}}, \tag{5.27}$$

and $\chi_4-\chi_6$ are roots of the cubic equation in ζ

$$\begin{vmatrix} (G_6-\beta\zeta^2) & \eta_1\eta_2 k_{16} & \eta_1 k_{16} \\ \eta_1\eta_2 k_{16} & (G_8-\beta\zeta^2) & \eta_2 k_{16} \\ \eta_1 k_{16} & \eta_2 k_{16} & (k_{16}-\alpha_1\zeta^2) \end{vmatrix} = 0. \tag{5.28}$$

After some manipulation it can be shown that the lowest curve in the ξ-ω plane is tangent to the ξ-axis at $\xi=\phi$. The set of curves in the ξ-ω plane given by (5.20) are shown in Figure 3.

The group of equations governing the secondary motion again reduce to (4.27) and (4.30) so that the corresponding curves are the same as those obtained in section 4. These curves are expected to lie above the six curves obtained from the main group of equations in Figure 3. Again it is possible to substitute back into the equations of section 2 to obtain more accurate values.

REFERENCES

1. A.E. Green, P.M. Naghdi and M.L. Wenner: On the theory of rods 1. Derivations from the three-dimensional equations, Proc. Roy. Soc. A 337, 451-483, 1974.

2. A.E. Green, P.M. Naghdi and M.L. Wenner: On the theory of rods II. Developments by direct approach, Proc. Roy. Soc. A 337, 485-507, 1974.

3. A.E. Green and N. Laws: A general theory of rods, Proc. Roy. Soc. A 293, 145-155, 1966.

4. A.E. Green, R.J. Knops and N. Laws: Large deformations, superposed small deformations and stability of elastic rods, Int. J. Solids Structures 4, 555-577, 1968.

5. A.E.H. Love: A treatise on the mathematical theory of elasticity, New York: Dover Publications, 1944.

6. G. Eason: Wave propagation in a naturally curved elastic rod, J. Sound and Vibration 36, 491-511, 1974.

7. A.E. Green, N. Laws and P.M. Naghdi: A linear theory of straight elastic rods, Archs. Rat. Mech. Anal. 25, 285-298, 1967.

Professor G. Eason,
Department of Mathematics,
University of Strathclyde,
Glasgow. Scotland.

The Kelvin Neumann problem

F. URSELL

1. INTRODUCTION

Consider a ship in steady uniform motion on the free surface of a fluid. It is well known that the ship is accompanied by a wave pattern which is steady relative to the ship; the calculation of this wave pattern and of the associated wave resistance is a central problem of ship hydrodynamics which however remains largely unsolved even although drastic simplifying assumptions have been made. Thus viscosity is often neglected; then the fluid motion is irrotational and can be described by a velocity potential, the stress in the fluid is a pressure which can reasonably be assumed to be constant at the free surface, and the velocity normal to the ship's hull and the velocity normal to the free surface both vanish. Even this simplified problem cannot be solved. One difficulty is that the free surface is not prescribed but must be determined during the solution of the problem. Thus additional simplifying assumptions are usually made. Many calculations are concerned with thin-ship theory: the ship is assumed to be so thin that the fluid motion relative to the ship is nearly a uniform stream. A perturbation procedure can then be set up in terms of a small thickness-parameter. The first approximation is linear and can be treated mathematically, but its range of validity is not adequate for many applications, and various non-linear approximations have therefore been studied. For instance, in some work the full non-linear condition on the ship's hull has been used together with the linearized free-surface condition on the mean free surface. The resulting problem is the so-called Kelvin-Neumann problem with which the present work is concerned. This approximation can be justified when the translating body is deeply submerged, but in recent work it has also been used for bodies intersecting the free surface. It is well understood that the approximation is then inconsistent even for second-order thin-ship theory since only some of the second-order terms are retained while others are omitted. The corrections that are obtained are therefore of doubtful practical value but may perhaps turn out to be in the right direction. It is hoped also that the experience gained in this way may be of use in treating the full inviscid problem in the future.

In the present work, however, we shall not be concerned with the practical applicability or logical consistency of the Kelvin-Neumann formulation for surface-piercing bodies but merely with certain mathematical aspects of the two-dimensional Kelvin-Neumann problem. The free surface of the fluid is then represented by a horizontal straight line, and the body is represented by a curve intersecting this line. It will also be supposed that the curve intersects the line at right angles. (According to the full theory, the corners would be stagnation points where the perturbation velocity is equal to the forward speed and where any perturbation scheme must fail, but we are ignoring this difficulty.) It has been suggested that any solution of the problem must have singularities in the corners. We shall see, however, that such singularities need not occur if at infinity the velocity potential is allowed to tend to infinity logarithmically with distance. The corresponding velocity components remain finite in the corners and at infinity.

2. STATEMENT OF THE MATHEMATICAL PROBLEM

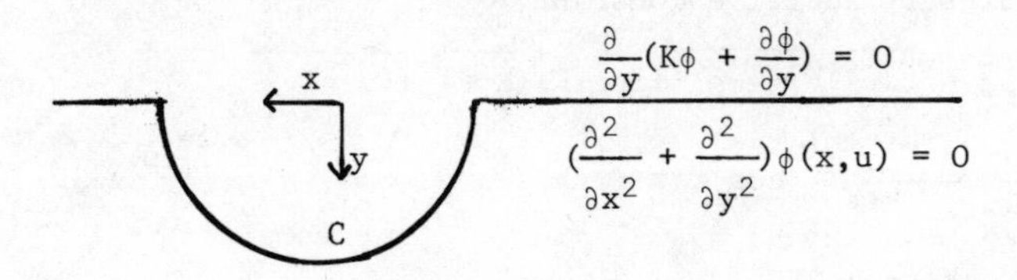

A cylinder, with its generators horizontal, is moving along the free surface of the fluid with a constant velocity $-U$ which is horizontal and normal to the generators. Take coordinate axes moving with the cylinder, the x-axis horizontal and the y-axis vertical, y increasing with depth. We shall be concerned with motions which are steady relative to the cylinder. The velocity potential is then written in the form $Ux + \phi(x,y)$.

Since the density is assumed uniform we have

$$\left(\frac{\partial^2}{\partial x^2} + \frac{\partial^2}{\partial y^2}\right)\phi(x,y) = 0 \text{ in the fluid.} \tag{2.1}$$

Since there is no flow across the boundary C of the cylinder we have

$$\frac{\partial\phi}{\partial n} = -U\frac{\partial x}{\partial n} \text{ on } C \tag{2.2}$$

where $\partial/\partial n$ denotes differentiation normal to C.

Since there are no waves formed upstream of the cylinder we have

$$\frac{\partial\phi}{\partial x} \to 0 \text{ and } \frac{\partial\phi}{\partial y} \to 0 \text{ when } x \to -\infty. \tag{2.3}$$

As was stated in the introduction, the curve C is assumed to intersect the line $y = 0$ at right angles. Near the corners, linearization cannot be justified. Let it nevertheless be applied along the whole of the free surface. Then the Kelvin-Neumann free-surface condition is

$$\frac{\partial}{\partial y}\left(K\phi + \frac{\partial}{\partial y}\right) = 0 \text{ on } y = 0 \text{ outside } C, \tag{2.4}$$

where $K = g/U^2$. (See e.g. Wehausen and Laitone, 1960, eqn. 10.15.)

3. THE SOURCE POTENTIAL

In subsequent work we shall need the potential of a source at (ξ,η). This will be denoted by $G(x,y;\ \xi,\eta)$; it also depends on K. It is known (Wehausen and Laitone, 1960, eqn. 13.44) that, except for an arbitrary additive constant,

$$G(x,y;\xi,\eta)=\log\{K\sqrt{(x-\xi)^2+(y-\eta)^2}+\log\{K\sqrt{(x-\xi)^2+(y+\eta)^2}\}$$
$$+2\int_0^\infty \frac{e^{-k(y+\eta)}}{k-K}\cos k(x-\xi)dk$$
$$-2\pi e^{-K(y+\eta)}\sin K(x-\xi) \tag{3.1}$$

where the symbol $⨍$ indicates that the Cauchy Principal Value of the integral is to be taken. Let us write

$$⨍_0^\infty \frac{e^{-kY}}{k-K} \cos kX \, dk = C(KX,\ KY).$$

Properties of this function are given in the appendix at the end of this paper. In particular,

$$C(K(x-\xi),\ K(y+\eta)) - \pi e^{-K(y+\eta)} \sin K(x-\xi) \to 0 \quad \text{when } x - \xi \to -\infty,$$

and it follows that there are no wave terms in (3.1) when $x - \xi \to -\infty$. Thus G( ) satisfies the radiation condition. Evidently $G(x,y;\ \xi,\eta)$ is not symmetric in (x,y) and (ξ,η). It should be noted that the source potential is logarithmically infinite at infinity, and this suggests that the solution of the boundary-value problem (see §2 above) may have the same property. This will in fact be shown in §4 below. It is important to note that G( ) is not an even function of $x-\xi$. It is also important to note that the surface singularity is replaced by a weaker singularity when $\eta \to 0$. The potential $G(x,y;\ \xi,0)$ will be described as a <u>weak surface singularity</u>. Similarly $\partial G(x,y;\ \xi,0)/\partial x$ (apparently a horizontal dipole) is a vortex. These results follow from the expansions in the appendix.

4. THE LEAST SINGULAR SOLUTION FOR THE HALF-IMMERSED SEMI-CIRCLE

Consider the potential $\phi(x,y) = \phi(r \sin\theta,\ r\cos\theta)$, given by the multipole expansion

$$\frac{\phi(x,y)}{Ua} = AP(Kx,\ Ky) + BQ(Kx,\ Ky)$$

$$+\alpha_0 a\left(\frac{\cos\theta}{r} - K\log Kr\right) + \sum_{m=1}^{\infty} \alpha_m a^{2m+1}\left(\frac{\cos(2m+1)\theta}{r^{2m+1}} + \frac{K}{2m}\frac{\cos 2m\theta}{r^{2m}}\right)$$

$$+ \sum_{m=0}^{\infty} \beta_m a^{2m+2}\left(\frac{\sin(2m+2)\theta}{r^{2m+1}} + \frac{K}{2m+1}\frac{\sin(2m+1)\theta}{r^{2m+1}}\right), \tag{4.1}$$

where

$$P(Kx,Ky) = \{\log Kr + ⨍_0^\infty e^{-ky} \cos kx \frac{dk}{k-K}\} + \{\{-\pi e^{-Ky}\sin Kx\}\},$$

$$= \{P_1(Kx,Ky)\} + \{\{P_2(Kx,Ky)\}\}, \text{ say,}$$

and $Q(Kx,Ky) = \{\fint_0^\infty e^{-ky} \sin kx \frac{dk}{k-K}\} + \{\{\pi e^{-Ky} \cos Kx\}\},$

$= \{Q_1(Kx,Ky)\} + \{\{Q_2(Kx,Ky)\}\}$, say,

where the Cauchy Principal Value is to be taken at $k = K$. (An arbitrary constant may be added to $\phi(x,y)$. It is readily verified that each term of the right-hand side of (4.1) satisfies Laplace's equation (2.1) and the free-surface condition (2.4) and the radiation condition (2.3). The boundary condition

$$\left<\frac{\partial\phi}{\partial r}\right>_{r=a} = -U \sin\theta, \; -\tfrac{1}{2}\pi \leqslant \theta \leqslant \tfrac{1}{2}\pi, \tag{4.2}$$

on the semi-circle C is also satisfied if the coefficients A, B, α_m, β_m are chosen so that in the interval $-\frac{1}{2}\pi < \theta < \frac{1}{2}\pi$ the equations

$$\begin{aligned} &A\left<a \frac{\partial}{\partial r} P_1\right> + B\left<a \frac{\partial}{\partial r} Q_2\right> \\ &-\alpha_o(\cos\theta - Ka) - \sum_{m=1}^{\infty} (2m+1)\alpha_m\left(\cos(2m+1)\theta + \frac{Ka}{2m+1}\cos 2m\theta\right) = 0 \end{aligned} \tag{4.3}$$

and

$$\begin{aligned} &A\left<a \frac{\partial}{\partial r} P_2\right> + B\left<a \frac{\partial}{\partial r} Q_1\right> + \sin\theta \\ &- \sum_{m=o}^{\infty} (2m+2)\beta_m\left(\sin(2m+2)\theta + \frac{Ka}{2m+2}\sin(2m+1)\theta\right) = 0 \end{aligned} \tag{4.4}$$

are satisfied. (Here and elsewhere brackets $< >$ are used to indicate that r is put equal to a inside the brackets.) The equations (4.3) and (4.4) are series expansions resembling Fourier series. When A and B have been chosen arbitrarily, these equations can be shown to have unique solutions α_m, β_m ($m = 0,1,2, \ldots$) except possibly at a discrete set of irregular values of Ka, and it can also be shown that $\alpha_m = O(1/m^2)$ and $\beta_m = O(1/m^2)$ by arguments used in the proof of Theorem 1 below. (The details are omitted.) The series for the two velocity components are thus seen to converge only slowly; at $\theta = \pm\frac{1}{2}\pi$ the horizontal velocity is in general discontinuous and the vertical velocity is logarithmically infinite.

We now obtain our principal result.

Theorem 1. The coefficients A and B in (4.1) can be chosen

uniquely so that $\alpha_m = O(1/m^3)$ and $\beta_m = O(1/m^3)$. The corresponding solution (the least singular solution) has velocities which are continuous and bounded in the corners. (Actually it will be seen that $\alpha_m = O(1/m^4)$ and $\beta_m = O(1/m^4)$.)

Proof of Theorem 1. If such a rapidly converging solution exists then we may put $\theta = \frac{1}{2}\pi$ in (4.3) and obtain

$$A\langle a \frac{\partial}{\partial r} P_1\rangle_{\frac{1}{2}\pi} + B\langle a \frac{\partial}{\partial r} Q_2\rangle_{\frac{1}{2}\pi} - Ka\sum_0^\infty \alpha_m \cos m\pi = 0 \tag{4.5}$$

On combining this with (4.3) we see that

$$A\{\langle a \frac{\partial}{\partial r} P_1\rangle - \langle a \frac{\partial}{\partial r} P_1\rangle_{\frac{1}{2}\pi}\} + B\{\langle a \frac{\partial}{\partial r} Q_2\rangle - \langle a \frac{\partial}{\partial r} Q_2\rangle_{\frac{1}{2}\pi}\}$$

$$= \sum_0^\infty (2m+1)\alpha_m\{\cos(2m+1)\theta + \frac{Ka}{2m+1}(\cos 2m\theta - \cos m\pi)\}. \tag{4.6}$$

Similarly, on putting $\theta = \frac{1}{2}\pi$ in (4.4) we find that

$$A\langle a \frac{\partial}{\partial r} P_2\rangle_{\frac{1}{2}\pi} + B\langle a \frac{\partial}{\partial r} Q_1\rangle_{\frac{1}{2}\pi} - Ka\sum_0^\infty {}_m \sin(m+\tfrac{1}{2})\pi = -1. \tag{4.7}$$

On combining this with (4.4) we see that

$$A\{\langle a \frac{\partial}{\partial r} P_2\rangle - \sin\theta\langle a \frac{\partial}{\partial r} P_2\rangle_{\frac{1}{2}\pi}\} + B\{\langle a \frac{\partial}{\partial r} Q_1\rangle - \sin\theta\langle a \frac{\partial}{\partial r} Q_1\rangle_{\frac{1}{2}\pi}\}$$

$$= \sum_0^\infty (2m+2)\beta_m\{\sin(2m+2)\theta + \frac{Ka}{2m+2}(\sin(2m+1)\theta - \sin\theta\sin(m+\tfrac{1}{2})\pi)\} \tag{4.8}$$

We shall now verify that the solutions of (4.6) and (4.8) satisfy $\alpha_m = O(1/m^4)$ and $\beta_m = O(1/m^4)$, by transforming (4.6) and (4.8) into systems of simultaneous equations. Consider e.g. (4.8). Multiply in turn by the complete set $\sin(2\ell+2)$, $(\ell = 0,1,2, \ldots ,)$ and integrate over $(-\frac{1}{2}\pi, \frac{1}{2}\pi)$. An infinite set of equations of the form

$$(2\ell+2)\beta_\ell + Ka\sum_{m=0}^\infty a_{\ell m}(2m+2)\beta_m = c_\ell, \quad = 0,1,2, \ldots, \tag{4.9}$$

is obtained where c_ℓ is the Fourier coefficient of the left-hand side of (4.8) and where the coefficients $a_{\ell m}$ can be found explicitly by elementary integrations. Write (4.9) in the form

$$(2\ell+2)^3\beta_\ell + Ka\sum_{m=0}^\infty \left(\frac{2\ell+2}{2m+2}\right) a_{\ell m}(2m+2)^3\beta_m = (2\ell+2)^2 c_\ell. \tag{4.10}$$

By integration by parts it can be shown that $c_\ell = O(1/\ell^3)$, thus

$\Sigma(2\ell+2)^4 c_\ell^2$ is convergent; and it can also be shown that $\sum_m\sum\left(\frac{2\ell+2}{2m+2}\right)^4 a_{\ell m}^2$ is convergent. The theory of the system (4.10) is therefore analogous to the theory of Fredholm integral equations of the second kind. It follows that (except possibly at a set of irregular values of Ka) there is a unique solution (β_ℓ) such that $\Sigma(2\ell+2)^6\beta_\ell^2$ is convergent. To show that $\beta_\ell = O(1/\ell^4)$, note from (4.10) that

$$|(2\ell+2)^3\beta_\ell-(2\ell+2)^2c_\ell|^2=(Ka)^2|\sum_{m=0}^{\infty}\left(\frac{2\ell+2}{2m+2}\right)^2 a_{\ell m}(2m+2)^3\beta_m|^2$$

$$\leqslant (Ka)^2\left(\sum_{m=0}^{\infty}\left(\frac{2\ell+2}{2m+2}\right)^4 a_{\ell m}^2\right)\left(\sum_{m=0}^{\infty}(2m+2)^6\beta_m^2\right).$$

by Cauchy's inequality,

$$\leqslant \text{const.} \sum_{m=0}^{\infty}\left(\frac{2\ell+2}{2m+2}\right)^4 a_{\ell m}^2.$$

It can be shown that $\sum_{m=0}^{\infty}\left(\frac{2\ell+2}{2m+2}\right)^4 a_{\ell m}^2 = O(1/\ell^2)$, and it follows that $\beta_\ell = O(1/\ell^4)$. Evidently, from (4.8), β_ℓ is of the form

$$\beta_\ell = A\beta_\ell(P_2) + B\beta_\ell(Q_1), \tag{4.11}$$

in an obvious notation, where $\beta_\ell(P_2)$ and $\beta_\ell(Q_1)$ are independent of A and B, and where $\beta_\ell(P_2) = O(1/\ell^4)$ and $\beta_\ell(Q_1) = O(1/\rho^4)$. Thus, from (4.7),

$$\begin{aligned} &A\{\langle a\frac{\partial}{\partial r}P_2\rangle_{\frac{1}{2}\pi} - Ka\sum_0^\infty \beta_m(P_2)\sin(m+\tfrac{1}{2})\pi\} \\ &+ B\{\langle a\frac{\partial}{\partial r}Q_1\rangle_{\frac{1}{2}\pi} - Ka\sum_0^\infty \beta_m(Q_1)\sin(m+\tfrac{1}{2})\pi\} = -1. \end{aligned} \tag{4.12}$$

Similarly, from (4.6)

$$a_\ell = A\alpha_\ell(P_1) + B\alpha_\ell(Q_2), \tag{4.13}$$

where $\alpha_\ell(P_1) = O(1/\ell^4)$ and $\alpha_\ell(Q_2) = O(1/\ell^4)$, thus, from (4.5),

$$\begin{aligned} &A\{\langle a\frac{\partial}{\partial r}P_1\rangle_{\frac{1}{2}\pi} - Ka\sum_0^\infty \alpha_m(P_1)\cos m\pi\} \\ &+ B\{\langle a\frac{\partial}{\partial r}Q_2\rangle_{\frac{1}{2}\pi} - Ka\sum_0^\infty \alpha_m(Q_2)\cos m\pi\} = 0. \end{aligned} \tag{4.14}$$

From (4.12) and (4.14) the values of A and B are uniquely determined except when the determinant of the system (4.12), (4.14) vanishes; hence α_ℓ and β_ℓ are determined from (4.13) and (4.11); hence $\phi(x,y)$ is determined from (4.1) except for an arbitrary additive constant. It is obvious that the potential $\phi(x,y)$ does indeed satisfy the boundary condition (4.2). It remains to be verified that the determinant of the system (4.12), (4.14) does not vanish identically; this can be done by considering the system for small Ka. Thus the procedure described above is effective except possibly at a discrete set of values of Ka and provides a construction for the least singular potential $\phi(x,y)$. It can be shown that for general values of Ka the logarithmic terms in (4.1) do not cancel, thus this potential is logarithmically infinite at infinity.

This concludes the proof of Theorem 1.

5. REPRESENTATION BY SOURCES

The method of multipoles, described in §4 above, is appropriate for the half-immersed semi-circle; for other cross-sections it can be generalized by conformal mapping (see e.g. Ursell 1949 for the corresponding construction of wave-free potentials at zero mean speed). An alternative approach uses distributions of wave sources over the boundary curve C but additional terms are needed at the ends. Thus for the half-immersed semi-circle we find

Theorem 2. The least singular solution (obtained in Theorem 1 above) can be represented in the form

$$\phi(x,y) = U\int_{-\frac{1}{2}\pi}^{\frac{1}{2}\pi} \mu(\theta)G(x,y;a\sin\theta,a\cos\theta)a\,d\theta \tag{5.1}$$

$$-\frac{U\mu(\frac{1}{2}\pi)}{K}G(x,y;a,0) \tag{5.2}$$

$$-\frac{U\mu(-\frac{1}{2}\pi)}{K}G(x,y;-a,0), \tag{5.3}$$

where the coefficients $\mu(\frac{1}{2}\pi)$ in (5.2) and $\mu(-\frac{1}{2}\pi)$ in (5.3) are the limits of the source-density function $\mu(\theta)$ when $\theta \to \pm\frac{1}{2}\pi$. A corresponding result holds for the least singular solution exterior to an arbitrary boundary curve C intersecting the horizontal at right angles.

To motivate the following proof let us recall the corresponding results in potential theory and in acoustics (Lamb 1932, §58 and §290). These are obtained from Green's theorems which states that the exterior solution $\phi(x,y)$ can be represented by sources of density $(2\pi)^{-1}\partial\phi/\partial n$ and normal dipoles of density $-(2\pi)^{-1}\phi$ distributed over C. Consider now the interior potential ϕ_{int} such that $\phi_{int} = \phi$ on C; let the corresponding normal velocity on C be denoted by $\partial\phi_{int}/\partial n$. Then, by Green's theorem, sources of density $(2\pi)^{-1}\partial\phi_{int}/\partial n$ and normal dipoles of density $-(2\pi)^{-1}\phi_{int} = -(2\pi)^{-1}\phi$, distributed over C, generate a null field exterior to C. By subtraction it follows that the exterior solution $\phi(x,y)$ can be represented by a source distribution of density $(2\pi)^{-1}\partial(\phi-\phi_{int})/\partial n$ over C. Similar ideas will now be applied to the Kelvin-Neumann problem.

<u>Proof of Theorem 2.</u> (Not all the details will be given.) Apply Green's theorem to the least singular solution $\phi(x,y)$ and to the reversed source function $\bar{G}(x,y;\xi,\eta) = G(\xi,\eta;x,y)$ in the region bounded by the contour C, by a large semi-circle S(R), and by the two segments of the x-axis between C and S(R). (Note that ϕ and $\bar{G}$ involve arbitrary additive constants.) It is assumed that (ξ,η) lies in this region. Then we find that

$$-2\pi\phi(\xi,\eta) = \int_C \{\phi(x,y)\frac{\partial}{\partial n}\bar{G}(x,y;\xi,\eta)$$

$$-\bar{G}(x,y;\xi,\eta)\frac{\partial}{\partial n}\phi(x,y)\}ds(x,y)$$

$$+\int_{S(R)}\{\quad\}ds(x,y)$$

$$+\int_a^R\left(\phi\frac{\partial\bar{G}}{\partial y}-\bar{G}\frac{\partial\phi}{\partial y}\right)_{y=0}dx+\int_{-R}^{-a}\left(\frac{\partial\bar{G}}{\partial y}-\bar{G}\frac{\partial\phi}{\partial y}\right)_{y=0}dx.$$

Also

$$\int_a^R\left(\phi\frac{\partial\bar{G}}{\partial y}-\bar{G}\frac{\partial\phi}{\partial y}\right)dx$$

$$= -\frac{1}{K}\int_a^R\left(\phi\frac{\partial^2\bar{G}}{\partial y^2}-\bar{G}\frac{\partial^2\phi}{\partial y^2}\right)dx \text{ from the free-surface condition}$$

$$= \frac{1}{K}\int_a^R\left(\phi\frac{\partial^2\bar{G}}{\partial x^2}-\bar{G}\frac{\partial^2\phi}{\partial x^2}\right)dx \text{ from Laplace's equation,}$$

$$= \frac{1}{K}\left[\phi \frac{\partial \bar{G}}{\partial x} - \bar{G} \frac{\partial \phi}{\partial x}\right]_{(a,0)}^{(R,0)} \tag{5.4}$$

and similarly

$$\int_{-R}^{-a} (\phi \frac{\partial \bar{G}}{\partial y} - \bar{G} \frac{\partial \phi}{\partial y}) dx = \frac{1}{K}\left[\phi \frac{\partial \bar{G}}{\partial x} - \bar{G} \frac{\partial \phi}{\partial x}\right]_{(-R,0)}^{(-a,0)}$$

Thus

$$-2\pi\phi(\xi,\eta) = \int_C \{\phi \frac{\partial \bar{G}}{\partial n} - \bar{G} \frac{\partial \phi}{\partial n}\} ds + \frac{1}{K}\left[\phi \frac{\partial \bar{G}}{\partial x} - \bar{G} \frac{\partial \phi}{\partial x}\right]_{(a,0)}^{(-a,0)}$$

$$+ \int_{S(R)} \{ \quad \} ds - \frac{1}{K}\left[\phi \frac{\partial \bar{G}}{\partial x} - \bar{G} \frac{\partial \phi}{\partial x}\right]_{(R,0)}^{(-R,0)} \quad . \tag{5.5}$$

Let $\phi_{int}(x,y)$ denote the least singular interior potential satisfying $\phi_{int} = \phi$ on C. (The construction of this potential involves the superposition of regular wave potentials and resembles the construction of the least singular exterior potential. The details are omitted.) Apply Green's theorem to ϕ_{int} and to $\bar{G}$ in the interior region bounded by C and by the x-axis. Then

$$0 = \int_C \{\phi_{int} \frac{\partial \bar{G}}{\partial n} - \bar{G} \frac{\partial}{\partial n} \phi_{int}\} ds$$

$$- \int_{-a}^{a} (\phi_{int} \frac{\partial \bar{G}}{\partial y} - \bar{G} \frac{\partial}{\partial y} \phi_{int}) dx$$

$$= \int_C \{ \ \} ds - \left[\phi_{int} \frac{\partial \bar{G}}{\partial x} - \bar{G} \frac{\partial}{\partial x} \phi_{int}\right]_{(-a,0)}^{(a,0)} \tag{5.6}$$

as in (5.5) above. On subtracting (5.6) from (5.5) we find that

$$-2\eta\phi(\xi,\eta) = -\int_C \frac{\partial}{\partial n}(\phi - \phi_{int}) \bar{G}(x,y;\xi,\eta) ds(x,y)$$

$$+ \frac{1}{K}\left[\frac{\partial}{\partial x}(\phi - \phi_{int}) \bar{G}\right]_{(a,0)}^{(-a,0)}$$

$$+ \int_{S(R)} \{ \quad \} ds - \frac{1}{K}\left[\phi \frac{\partial \bar{G}}{\partial x} - \bar{G} \frac{\partial \phi}{\partial x}\right]_{(R,0)}^{(-R,0)} \quad . \tag{5.7}$$

Write

$$\mu(\theta) = \frac{2\pi}{U} \frac{\partial}{\partial n}(\phi - \phi_{int}) \ . \tag{5.8}$$

where the normal gradients are to be evaluated at $(x,y) = (a \sin\theta, a \cos\theta)$; also note that $\bar{G}(x,y;\xi,\eta) = G(\xi,\eta;x,y)$. Then it is seen that (5.7) is equivalent to Theorem 2, except for the two terms arising from the large semi-circle. It can be shown that these contribute at most an additive constant. This concludes the proof of Theorem 2.

Note that for reasons of brevity we have omitted proofs that the least singular interior potential exists, and that the large semi-circle contributes at most a constant to the potential.

We can now see how the least singular potential can be constructed by means of an integral equation. Write

$$\phi(x,y) = U\int_{-\frac{1}{2}\pi}^{\frac{1}{2}\pi} \mu(\theta)G(x,y;a \sin\theta, a\cos\theta)a\, d\theta$$

$$- UapG(x,y;a,0) - UaqG(x,y;-a,0),$$

where the last two terms (apparently wave sources) are actually weak surface singularities. The function $\mu(\theta)$ and the constants p and q are to be determined. On applying the boundary condition (4.2) we find that

$$\pi\mu(\alpha)+\int_{-\frac{1}{2}\eta}^{\frac{1}{2}\eta} \mu(\theta)\langle a\frac{\partial}{\partial r}G(r \sin\alpha, r\cos\alpha; a\sin\theta, a\cos\theta)\rangle d\theta$$

$$= p\langle a\frac{\partial}{\partial r} G(r\sin\alpha, r\cos\alpha; a,0)\rangle$$

$$+ q\langle a\frac{\partial}{\partial r} G(r\sin\alpha, r\cos\alpha; -a,0)\rangle$$

$$- \sin\alpha. \tag{5.9}$$

This is a Fredholm equation of the second kind, and (except possibly at a discrete set of irregular values of Ka) there is a unique solution which is evidently of the form

$$\mu(\theta) = p\mu_+(\theta) + q\mu_-(\theta) + \mu_o(\theta), \tag{5.10}$$

where the functions $\mu_+(\theta)$, $\mu_-(\theta)$, $\mu_o(\theta)$ are the solutions corresponding to the three known functions on the right-hand side of (5.9). From Theorem 2 we have $p = \mu(\frac{1}{2}\pi)/Ka$ and $q = \mu(-\frac{1}{2}\pi)/Ka$; thus, on putting $\theta = \frac{1}{2}\pi$ and $\theta = -\frac{1}{2}\pi$ in (5.10) we see that

$$Kap = p\mu_+(\tfrac{1}{2}\pi) + q\mu_-(\tfrac{1}{2}\pi) + \mu_o(\tfrac{1}{2}\pi),$$

and

$$Kaq = p\mu_+(-\tfrac{1}{2}\pi) + q\mu_-(-\tfrac{1}{2}\pi) + \mu_o(-\tfrac{1}{2}\pi).$$

From these equations the two constants p and q can be found, and the source density is then given uniquely by (5.10).

If p and q are instead given arbitrary values, then the solution (5.10) can be shown to have unbounded vertical velocities in the corners. This corresponds to the solution (4.1) above when A and B are given arbitrary values.

6. DISCUSSION

It has been shown that for the half-immersed semi-circle the Kelvin-Neumann problem has a two-parameter set of solutions if the singularities in the two corners are at most weak surface singularities. There is just one solution, the least singular solution, for which the velocity is bounded in both corners. It has been shown how this solution can be constructed, either by the method of multipoles or by a distribution of wave sources over the boundary. (The source density satisfies an integral equation (5.9) involving additional end-contributions.) There is however no obvious physical reason why the condition of boundedness should be imposed in the corners. In the physical problem the perturbation velocity is in fact not small in the corners, and the linearization is not valid there. It would thus be equally reasonable to look for solutions of the perturbation equations which have weak or strong singularities in the corners but there is then no obvious way of deciding what singularities would be appropriate.

In an earlier unpublished version (1978) of this work the representation (5.10) was used with p = q = 0, and it was shown that the vertical velocity in the corners is then unbounded. This choice is arbitrary, as was pointed out to me by Mr. Katsuo Suzuki, whose criticism led me to a more thorough study of the problem. Suzuki's own choice of p and q is based on additional physical conditions and does not lead to the least singular solution. Although there is no obvious physical reason why the least singular solution should be preferred in two dimensions there may well be physical reasons why a corresponding boundedness condition should be applied in three dimensions, and this is one of the motivations of the present study.

The perturbation potential $\phi(x,y)$ becomes logarithmically unbounded at infinity; the velocities due to the logarithmic terms tend to 0 at infinity. There appears to be no physical reason for excluding such a solution; it will be recalled that the total potential is linearly unbounded at infinity.

It is sometimes stated that the perturbation potential must have strong singularities in the corners; it has been seen in the present work that this conclusion is incorrect. The usual argument is based on (5.5) above which from the corner at (a,0) contributes terms.

$$(2\pi K)^{-1}\phi(a,0)\ \frac{\partial}{\partial\xi}\ G(\xi,\eta;a,0)+(2\eta K)^{-1}\ \frac{\partial\phi}{\partial x}\ (a,0)G(\xi,\eta;a,0)$$

to the potential $\phi(\xi,\eta)$. Since $G(\xi,\eta;X,Y)$ represents a source at (X,Y) when $Y > 0$ this is often interpreted as the sum of a dipole term and a source term. This interpretation is in any case inconsistent since the dipole strength $\phi(a,0)$ and the source strength $\partial\phi(a,0)/\partial x$ would not be finite if there were strong singularities in the corners. Actually $G(\xi,\eta;a,0)$ is a weak surface singularity, and $\partial G(\xi,\eta;a,0)/\partial\xi$ is a vortex as can be seen from the expansions in the appendix but even these weaker singularities are not present in the least singular solution. The correct interpretation can be inferred from the representation (5.1 - 5.3) of the least singular solution, which consists of a continuous distribution of sources together with a discrete weak surface singularity in each corner. Near each corner the end effect of the source distribution is like a weak surface singularity and is cancelled by the corresponding discrete weak surface singularity term. Similarly near a corner the end effect of the dipole distribution is like a vortex and is cancelled by the corresponding discrete vortex term.

7. ACKNOWLEDGEMENT

I am grateful to Mr. Katsuo Suzuki for his valuable comments on an earlier version of this paper.

APPENDIX

Power series expansions associated with the source potential.

We consider the functions

$$C(Kx,Ky) = \int_0^\infty e^{-ky} \cos Kx \frac{dk}{k-K} \tag{A.1}$$

and

$$S(Kx,Ky) = \int_0^\infty e^{-ky} \sin kx \frac{dk}{k-K} \tag{A.2}$$

where the Cauchy Principal Value is to be taken at k = K. Write

$F(\zeta) = \int_0^\infty e^{-k\zeta} \frac{dk}{k-K}$ where ζ has positive real part. Then

$$\int_0^\infty e^{-p\zeta} F(\zeta) d\zeta = \int_0^\infty \frac{dk}{(k-K)} \frac{1}{(k+p)}, \text{ by changing the order of integration,}$$

$$= \frac{1}{p+K}\left[\log\left|\frac{k-K}{k+p}\right|\right]_{k=0}^{k=\infty} = \frac{1}{k+p} \log \frac{p}{k}$$

$$= \sum_{m=0}^{\infty} \frac{(-K)^m}{p^{m+1}} \log \frac{p}{K} . \tag{A.3}$$

Also, from Euler's definition of the gamma function,

$$\int_0 e^{-p} \frac{(K\zeta)^\nu}{\Gamma(\nu+1)} d\zeta = \frac{K^\nu}{p^{\nu+1}} ,$$

whence by differentiation with respect to ν,

$$\int_0^\infty e^{-p\zeta} \left(\frac{\partial}{\partial\nu} \frac{(K\zeta)^\nu}{\Gamma(\nu+1)}\right) d\zeta = \frac{K^\nu}{p^{\nu+1}} \log \frac{K}{p} . \tag{A.4}$$

On comparing this with (A.3), we find that

$$F(\zeta) = \sum_{m=0}^{\infty} (-1)^{m+1} \left(\frac{\partial}{\partial\nu} \frac{(K\zeta)^\nu}{\Gamma(\nu+1)}\right)_{\nu=m}$$

$$= \sum_{m=0}^{\infty} (-1)^{m+1} \left\{\frac{(K\zeta)^m \log K\zeta}{\Gamma(m+1)} - \frac{(K\zeta)^m}{(\Gamma(m+1))^2} \Gamma'(m+1)\right\}$$

whence

$$\int_0^\infty e^{-k\zeta} \frac{dk}{k-K} = - \sum_{m=0}^{\infty} (-1)^m \frac{(K\zeta)^m}{m!} \left(\log K - \frac{\Gamma'(m+1)}{\Gamma(m+1)}\right) \tag{A.5}$$

Now put $\zeta = y + ix = re^{i\theta}$, then

$$\int_0^\infty e^{-ky}(\cos kx - i \sin kx)\, \frac{dk}{k-K}$$

$$= \sum_{m=0}^{\infty} (-1)^{m+1} \frac{(Kr)^m}{m!} (\cos m\theta + i \sin m\theta) \,.$$

$$. \left(\log Kr - \frac{\Gamma'(m+1)}{\Gamma(m+1)} + i\theta\right)$$

On taking real and imaginary parts we find that

$$C(Kx,Ky) = \int_0 e^{-ky}\cos kx \,\frac{dk}{k-K}$$

$$= \sum_0 (-1)^{m+1} \frac{(Kr)^m}{m!}\{\cos m\theta(\log Kr - \frac{\Gamma'(m+1)}{\Gamma(m+1)}) - \theta \sin m\theta\}, \qquad \text{(A.6)}$$

and that

$$S(Kx,Ky) = \int_0^\infty e^{-ky}\sin kx \,\frac{dk}{k-K}$$

$$= \sum_0^\infty (-1)^m \frac{(Kr)^m}{m!} \{\sin m\theta(\log Kr - \frac{\Gamma'(m+1)}{\Gamma(m+1)}) + \theta \cos m\theta\}. \qquad \text{(A.7)}$$

It is known that $\frac{\Gamma'(1)}{\Gamma(1)} = -\gamma$, and $\frac{\Gamma'(m+1)}{\Gamma(m+1)} = -\gamma + \frac{1}{1} + \frac{1}{2} + \ldots + \frac{1}{m}$, where $\gamma = 0.5772 \ldots$ is Euler's constant.

To find the behaviour of $F(y+ix)$ when $x \to +\infty$, consider

$$F(\zeta) = \int_0^\infty e^{-k\zeta} \frac{dk}{k-K}$$

$$= \int_0^\infty e^{-k\zeta} \frac{dk}{k-K} - \pi i e^{-K\zeta},$$

by adding and subtracting the integral along a small indentation below $k = K$,

$$= \int_0^{\infty \exp(-1/4 i\pi)} e^{-k\zeta} \frac{dk}{k-K} - \pi i e^{-K\zeta}, \qquad \text{(A.8)}$$

where the indented path along the real k-axis has been deformed into the path $\arg k = -\frac{1}{4}\pi$, on which the integrand is bounded. The integral in (A.8) is easily seen to tend to 0 when $Kx \to +\infty$, and we thus find that

$$F(\zeta) = C(Kx,Ky) - iS(Kx,Ky)$$

$$\sim -\pi i e^{-K(y+ix)} = -\pi i e^{-Ky}(\cos Kx - i \sin Kx)$$

$$= -\pi e^{-Ky}\sin Kx - \pi i e^{-Ky}\cos Kx \text{ when } Kx \to +\infty.$$

Since C( ) is an even function of Kx and S( ) is an odd function of Kx we infer that

$$C(Kx,Ky) - \pi e^{-Ky} \sin Kx \to 0 \text{ when } Kx \to -\infty, \tag{A.9}$$

and

$$S(Kx,Ky) + \pi e^{-Ky} \cos Kx \to 0 \text{ when } Kx \to -\infty. \tag{A.10}$$

We also need the expansion for the horizontal dipole:

$$\frac{\partial C}{\partial x} = - \int_0^\infty k e^{-ky} \sin kx \frac{dk}{k-K}$$

$$= - \int_0^\infty e^{-ky} \sin kx \, dk - K \int_0^\infty e^{-ky} \sin kx \frac{dk}{k-K}$$

$$= - \frac{x}{x^2+y^2} - KS(Kx,Ky)$$

$$= - \frac{\sin\theta}{r} - KS(Kx,Ky); \tag{A.11}$$

and the well-known expansions

$$e^{-Ky} \cos Kx = \sum_0^\infty (-1)^m \frac{(Kr)^m}{m!} \cos m\theta \tag{A.12}$$

$$e^{-Ky} \sin Kx = \sum_0^\infty (-1)^{m+1} \frac{(Kr)^m}{m!} \sin m\theta \tag{A.13}$$

REFERENCES

1. H. Lamb: Hydrodynamics, 6th Ed. Cambridge University Press, 1932.
2. F. Ursell: On the rolling motion of cylinders in the surface of a fluid, Quart. J. Mech. Appl. Math., 2, 335-353, 1949.
3. J. V. Wehausen and E. V. Laitone: Surface Waves, Handbuch der Physik, Vol. 9,Berlin: Springer, 446-778, 1960.

Professor F. Ursell,
Department of Mathematics,
University of Manchester,
Manchester, England.

Oscillations in a compressible laminar boundary layer

R. GRIBBEN

1. INTRODUCTION

We consider the flow in the laminar boundary layer of a compressible fluid when the main stream has the form,

$$U_1(x,t) = U_e(x)(1+\alpha \sin wt), \tag{1}$$

where x is distance along the surface, t is the time, $\alpha(<1)$ and w are constant and $U_e(x)$ is the mean velocity of the flow outside the boundary layer. Such flows have been studied previously by several authors (see e.g. [1] - [5]) and are of importance in, for example, flows over helicopter blades or about wing sections in flutter.

One difficulty with compressible boundary layer flows is that the external inviscid flow must be calculated in addition to that in the viscous boundary layer. We simplify this step here by assuming the Mach number M of the main stream to be small and the frequency parameter $s=wx/U_e$ to be large in such a way that $c = \alpha\gamma M^2 s/2 = O(1)$, where γ is the ratio of specific heats of the fluid (chosen to be 1.4 throughout). Note that if $c \to 0$ and temperature differences are small compared with the temperature the flow tends to the corresponding incompressible flow.

Two length scales, of widely differing magnitudes, can be defined for the boundary layer corresponding to an inner (Stokes) layer and a much thicker outer (Prandtl) layer. Associated dimensionless variables are used to write down the boundary layer equations in each layer. Solutions in the form of series in inverse powers of s can then be found in each of these layers and matched. It is crucial to the method's success that the leading order equations are linear. In fact, the enthalpy in the inner layer is obtained as a Fourier-like series in a distorted time variable and the wall heat transfer requires summation of such a series for various time values over a period. However, whilst for $c = \frac{1}{2}$, this task is accomplished without difficulty, as c increases towards its limiting value, $\gamma((\gamma-1)$ (= 3.5) the coefficients in the series decrease more slowly and for $c > 1$ accelerated convergence of the series is needed. A reasonably satisfactory way of doing this is described, and some results presented, in Section 6.

2. FORMULATION OF THE PROBLEM

The pressure p, density ρ and enthalpy I in the main stream satisfy

$$\frac{1}{\rho_1}\frac{\partial p_1}{\partial x} = -\left(\frac{\partial U_1}{\partial t} + U_1\frac{\partial U_1}{\partial x}\right), \quad \frac{\partial p_1}{\partial t} = \rho_1\left(\frac{\partial}{\partial t} + U_1\frac{\partial}{\partial x}\right)\left(I_1 + \frac{U_1^2}{2}\right),$$
$$p_1 = \rho_1 R I_1/c_p, \tag{2}$$

where suffix 1 refers to main stream values, c_p is the specific heat at constant pressure and R is the gas constant. With U_1 given by (1) solutions of (2) valid for small M and $c = O(1)$ are available in the form

$$p_1 = p_\infty E^{\gamma/(\gamma-1)} + O(M^2),\ \rho_1 = \rho_\infty E^{1/(\gamma-1)} + O(M^2),$$
$$I_1 = I_\infty E + O(M^2), \tag{3}$$

where $E = 1-(\gamma-1)c\cos wt/\gamma$ and suffix ∞ denotes reference conditions. We neglect the $O(M^2)$ terms in (3) when substituting for p_1, ρ_1 and I_1 in the boundary layer equations.

The latter can be written (see [1]) as

$$\frac{\partial^2\psi}{\partial t\partial Y} + \left\{\frac{\partial\psi}{\partial Y}, \psi\right\} = \left(\frac{\partial U_1}{\partial t} + U_1\frac{\partial U_1}{\partial x}\right)\frac{I}{I_1} + \frac{C\nu_\infty p_1}{p_\infty}\frac{\partial^3\psi}{\partial Y^3}, \tag{4a}$$

$$\frac{\partial}{\partial t}\left(\frac{I}{I_1}\right) + \left\{\frac{I}{I_1}, \psi\right\} = \frac{C\nu_\infty p_1}{p_\infty}\left[\frac{1}{\sigma}\frac{\partial^2}{\partial Y^2}\left(\frac{I}{I_1}\right) + \frac{1}{I_1}\left(\frac{\partial^2\psi}{\partial Y^2}\right)^2\right], \tag{4b}$$

where $\{a,b\} = \frac{\partial a}{\partial x}\cdot\frac{\partial b}{\partial y} - \frac{\partial a}{\partial y}\cdot\frac{\partial b}{\partial x}$, $Y = \frac{1}{\rho_\infty}\int_0^y \rho\, dy$ (y being the coordinate normal to the surface) and ψ is related to the velocity components u and v by $u = \partial\psi/\partial Y$, $v = -(\rho_\infty/\rho)(\partial\psi/\partial x + \partial Y/\partial t)$. Also ν is the kinematic viscosity, σ the Prandtl number and $C = (T_w/T_\infty)^{\frac{1}{2}}(T_\infty+T_s)/(T_w+T_s)$, the Sutherland constant, where T is the temperature, T_s a gas-dependent constant and suffix w refers to wall values.

In the inner layer we define dimensionless variables by $\psi = U_e(x)(2C\nu_\infty/w)^{\frac{1}{2}}f(x,\eta,\bar{t})$, $Y = (2C\nu_\infty/w)^{\frac{1}{2}}\eta$, $I/I_1 = h(x,\eta,\bar{t})$, $wt = \bar{t}$ and (4) become

$$s\left[\frac{\partial^2 f}{\partial\bar{t}\partial\eta} - \frac{p_1}{2p_\infty}\frac{\partial^3 f}{\partial\eta^3} - \alpha h\cos\bar{t}\right] + \frac{xU_e'}{U_e}\left[\left(\frac{\partial f}{\partial\eta}\right)^2 - f\frac{\partial^2 f}{\partial\eta^2} - h(1+\alpha\sin\bar{t})^2\right] + x\left\{\frac{\partial f}{\partial\eta}, f\right\} = 0, \tag{5a}$$

$$s\left[\frac{\partial h}{\partial\bar{t}} - \frac{p_1}{2\sigma p_\infty}\frac{\partial^2 h}{\partial\eta^2} - \frac{(\gamma-1)\rho_1 M^2}{2\rho_\infty}\left(\frac{\partial^2 f}{\partial\eta^2}\right)^2\right] - \frac{xU_e'}{U_e} f\frac{\partial h}{\partial\eta} + x\{h,f\} = 0, \tag{5b}$$

where $U_e' \equiv dU_e/dx$.

Similarly in the outer layer we write $\psi = (2C\nu_\infty xU_e)^{\frac{1}{2}}g(x,\xi,\bar{t})$, $Y = (2C\nu_\infty x/U_e)^{\frac{1}{2}}\xi$, $I/I_1 = k(x,\xi,\bar{t})$, when (4) become,

$$s\left[\frac{\partial^2 g}{\partial\bar{t}\partial\xi} - \alpha k\cos\bar{t}\right] + x\left\{\frac{\partial g}{\partial\xi}, g\right\} - \frac{g}{2}\frac{\partial^2 g}{\partial\xi^2} + \frac{xU_e'}{U_e}\left[\left(\frac{\partial g}{\partial\xi}\right)^2 - \frac{g}{2}\frac{\partial^2 g}{\partial\xi^2} - (1+\alpha\sin\bar{t})^2 k\right] - \frac{p_1}{2p_\infty}\frac{\partial^3 g}{\partial\xi^3} = 0, \tag{6a}$$

$$s\frac{\partial k}{\partial\bar{t}} + x\{k,g\} - \frac{g}{2}\frac{\partial k}{\partial\xi}\left(1 + \frac{xU_e'}{U_e}\right) - \frac{p_1}{2\sigma p_\infty}\frac{\partial^2 k}{\partial\xi^2} - \frac{(\gamma-1)\rho_1 M^2}{2\rho_\infty}\times\left(\frac{\partial^2 g}{\partial\xi^2}\right)^2 = 0. \tag{6b}$$

Note that $\eta = s^{\frac{1}{2}}\xi$.

The associated boundary conditions are

$$k \to 1, \quad \partial g/\partial\xi \to 1 + \alpha\sin\bar{t} \quad \text{as } \xi \to \infty, \tag{7}$$

$$\partial f/\partial x = \partial f/\partial\eta = 0, \; h = I_w/I_1 \quad \text{on} \quad \eta = 0, \tag{8}$$

where I_w = constant.

The aim is to solve (5) and (6) subject to (7) and (8) in inverse powers of s in the two layers and match the resulting solutions.

3. OUTER LAYER SOLUTIONS

With assumed series of the form

$$g = \sum_{m=0}^{\infty} s^{-m/2} g_m(x,\xi,\bar{t}),\ k = \sum_{m=0}^{\infty} s^{-m/2} k_m(x,\xi,\bar{t}),$$

we obtain on substitution into (6b),

$$k_o = K_o(x,\xi),\ g_o = \alpha \sin \bar{t} \int_0^{\xi} K_o d\xi + G_o(x,\xi),$$

and the equations satisfied by the time-independent functions, G_o and K_o, are obtained by considering the equations for g_2 and k_2. Thus, for example, k_2 satisfies

$$\frac{\partial k_2}{\partial \bar{t}} = \frac{g_o}{2} \frac{\partial k_o}{\partial \xi} \left(1 + \frac{xU'_e}{U_e}\right) + \frac{p_1}{2\sigma p_\infty} \frac{\partial^2 k_o}{\partial \xi^2} - x\{k_o, g_o\},$$

from (6b) and, hence, for k_2 to be 2π-periodic in $\bar{t}$, we require

$$\frac{\varepsilon}{2\sigma} \frac{\partial^2 K_o}{\partial \xi^2} + \tfrac{1}{2} \left(1 + \frac{xU'_e}{U_e}\right) G_o \frac{\partial K_o}{\partial \xi} - x\{K_o, G_o\} = 0, \tag{9}$$

where

$$\varepsilon(x) = \frac{1}{2\pi} \int_0^{2\pi} E^{\gamma/(\gamma-1)} d\bar{t}\ . \tag{10}$$

Similar considerations applied to the momentum equation (6a) yield the equation,

$$\frac{\varepsilon}{2} \frac{\partial^3 G_o}{\partial \xi^3} + \frac{G_o}{2} \frac{\partial^2 G_o}{\partial \xi^2} + \frac{xU'_e}{U_e} \left[\left(1 + \frac{\alpha^2}{2} (1-K_o)\right) K_o + \frac{G_o}{2} \frac{\partial^2 G_o}{\partial \xi^2} - \left(\frac{\partial G_o}{\partial \xi}\right)^2 \right] - x \left\{ \frac{\partial G_o}{\partial \xi}, G_o \right\} = 0. \tag{11}$$

The boundary conditions satisfied by G_o and K_o in (9) and (11) are

$$\lim_{\xi \to \infty} K_o = 1, \quad \lim_{\xi \to \infty} \partial G_o / \partial \xi = 1,$$

from (8) and the remaining ones, on $\xi = 0$, are obtained by matching with the inner solution.

4. INNER LAYER SOLUTIONS

Solutions of (5) in the form

$$f = \sum_{m=0}^{\infty} s^{-m/2} f_m(x,\eta,\bar{t}),\quad h = \sum_{m=0}^{\infty} s^{-m/2} h_m(x,\eta,\bar{t})$$

are sought. From (5b) h_o satisfies

$$\frac{\partial h_o}{\partial \bar{t}} - \frac{E^{\gamma/(\gamma-1)}}{2\sigma} \frac{\partial^2 h_o}{\partial \eta^2} = 0,$$

which becomes the one-dimensional diffusion equation,

$$\frac{\partial h_o}{\partial \theta} - \frac{1}{2\sigma}\frac{\partial^2 h_o}{\partial \eta^2} = 0, \tag{12}$$

on transforming $\bar{t}$ to θ defined by

$$d\theta = E^{\gamma/(\gamma-)} d\bar{t}, \quad \theta = 0 \text{ when } \bar{t} = 0,$$

so that $\theta(x, 2\pi) = 2\pi\varepsilon(x)$ (cf. (10)).

Equation (12) is solved by separation of variables in the form

$$h_o = A_{oo}(x)\eta + B_{oo}(x) + \sum_{n\neq 0} B_{on}(x)\exp(in\,\theta/\varepsilon - \eta\sqrt{2in\sigma/\varepsilon}), \tag{13}$$

where A_{oo}, B_{on} are arbitrary functions, on dropping exponentially large terms in η. In fact $A_{oo} = 0$ from matching with the outer solution and application of the boundary condition at $\eta = 0$ yields, from (8) and (13),

$$\frac{I_w}{I_\infty E} = \sum_n B_{on}\exp(in\,\theta/\varepsilon),$$

whence B_{on} is determined as

$$B_{on} = \frac{1}{2\pi\varepsilon}\frac{I_w}{I_\infty}\int_0^{2\pi} E^{1/(\gamma-1)}\exp(-in\theta(\bar{t})/\varepsilon)d\bar{t}.$$

Thus h_o is known from (13).

The first order solution h_1 is easy to calculate since wall conditions and equations are homogeneous to this order. We deduce $h_1 = A_{1o}\eta$ where $A_{1o} = \partial K_o/\partial\xi$ evaluated at $\xi = 0$.

Similar, though slightly more complicated, methods can be used to obtain the functions f_o and f_1 from (5a).

5. EXPRESSION FOR HEAT TRANSFER

The heat transfer per unit area from the wall to the gas is an important physical quantity and is defined by

$$q_w = -\left(k\frac{\partial T}{\partial y}\right)_{y_{oo}} = -\frac{I_1 p_1}{\sigma p_\infty}\left(\frac{C\mu_\infty \rho_\infty w}{2}\right)^{\frac{1}{2}}\{q^{(o)} + s^{-\frac{1}{2}}q^{(1)} + O(s^{-1})\},$$

where $q^{(j)} = (\partial h_j/\partial\eta)_{\eta=0}$ $(j = 0,1)$ and k is the thermal conductivity. Using the results of section 4 the dominant term $q^{(o)}$ is

$$q^{(o)} = -\sum_n (2in\sigma/\varepsilon)^{\frac{1}{2}} B_{on} e^{in\phi} = -\frac{I_w(\gamma-1)\sigma^{\frac{1}{2}}}{I_\infty\gamma}Q,$$

where

$$Q = \frac{2\gamma}{(\gamma-1)\varepsilon\sqrt{\varepsilon}} \sum_{n=1}^{\infty} \sqrt{n}\,\alpha_n \,(\cos n\phi - \sin n\phi), \tag{14}$$

$$\alpha_n = \frac{\varepsilon}{2\pi}\int_0^{2\pi} \frac{\cos n\phi \, d\phi}{E(\bar{t} : c)}, \tag{15}$$

and ϕ is related to $\bar{t}$ by $\varepsilon\, d\phi = d\theta = E^{\gamma/(\gamma-1)} d\bar{t}$, $\phi(o) = 0$. The question of summation of the series (14) for various values of c will be considered in section 6.

The first order term, $q^{(1)} = A_{1o} = (\partial K_o/\partial\xi)_{\xi=0}$ is obtained by solving the outer flow equations (9) and (11) for which the boundary conditions on $\xi = 0$, obtained by matching, are $G_o = \partial G_o/\partial\xi = 0$, $K_o = B_{oo}$. We shall be concerned here only with the dominant terms $q^{(o)}$.

Again, the skin friction can be obtained by similar methods.

6. SERIES SUMMATION

In summing the series in (14) for a range of values of ϕ (and hence $\bar{t}$) over a period in order, say, to draw a graph of the response of the dominant part of the heat transfer to time variations in the main stream, it is first necessary to evaluate the α_n numerically. Thus the relation between ϕ and $\bar{t}$ is integrated and stored and then used in the quadrature procedure in (15). As c is increased the integrand in (15) becomes badly behaved near $\phi = 0$ and 2π and it is better to use $\bar{t}$ as integration variable. A more serious difficulty, however, for $c > 1$ is that the series in (14) becomes more slowly convergent and some form of acceleration of the convergence is desirable. Two complementary methods are used.

First we apply 'summation by parts' ([6]), in which a Fourier series $S^*(\phi) = \sum_0^{\infty} a_n \exp(in\phi)$ is expressed in the alternative form

$$S^*(\phi) = \left\{ \sum_{n=1}^{\infty} T^*(a_n)e^{in\phi} + a_o(e^{-i\phi} - 2) + a_1 \right\} / 2(\cos\phi - 1), \tag{16}$$

where $T^*(a_n) = a_{n+1} + a_{n-1} - 2a_n$ is the second difference operator and is $o(a_n)$ as $n \to \infty$. Table 1 shows the application of this transformation to the series in (14) when $c = 2$ and $c = 3$ for

11 equally-spaced ϕ values and several numbers of terms of the series. The upper of the two entries in each case is the result of summing the series directly and the lower is the value obtained by applying a method of accelerated convergence. We see that the direct sums at the values of N quoted give no reliable indication

ϕ \ N	5	10	15	20
0	0.1710	0.3017	0.4121	0.5080
	0.7054	0.9845	1.127	1.212
$\pi/5$	-0.1297	-0.0265	-0.1143	-0.0377
	-0.0735	-0.0736	-0.0735	-0.0736
$2\pi/5$	-0.0148	-0.0207	-0.0246	-0.0276
	-0.0488	-0.0489	-0.0489	-0.0489
$3\pi/5$	-0.0622	-0.0166	-0.0549	-0.0216
	-0.0372	-0.0372	-0.0372	-0.0372
$4\pi/5$	-0.0097	-0.0131	-0.0153	-0.0170
	-0.0289	-0.0289	-0.0289	-0.0289
π	-0.0361	-0.0097	-0.0318	-0.0126
	-0.0216	-0.0216	-0.0216	-0.0216
$6\pi/5$	-0.0042	-0.0059	-0.0062	-0.0078
	-0.0138	-0.0138	-0.0138	-0.0138
$7\pi/5$	-0.0073	-0.0008	-0.0064	-0.0015
	-0.0039	-0.0038	-0.0038	-0.0038
$8\pi/5$	+0.0060	+0.0074	+0.0082	+0.0087
	+0.0127	+0.0126	+0.0125	+0.0125
$9\pi/5$	0.0870	0.0264	0.0764	0.0334
	0.0519	0.0538	0.0532	0.0535
2π	0.0710	0.3017	0.4121	0.5080
	0.7054	0.9845	1.127	1.212

(a) $c = 2$

ϕ \ N	5	10	15	20
0		0.0781	0.1092	0.1381
		0.3652	0.5124	0.6505
$\pi/5$		-0.0055	-0.0304	-0.0071
	-0.0184	-0.0184	-0.0184	-0.0184
$2\pi/5$		-0.0042	-0.0054	-0.0058
	-0.0125	-0.0125	-0.0125	-0.0125
$3\pi/5$		-0.0035	-0.0147	-0.0047
	-0.0096	-0.0096	-0.0096	-0.0096
$4\pi/5$		-0.0032	-0.0035	-0.0037
	-0.0074	-0.0075	-0.0075	-0.0075
π		-0.0023	-0.0086	-0.0028
	-0.0056	-0.0056	-0.0056	-0.0056
$6\pi/5$		-0.0012	-0.0013	-0.0014
	-0.0035	-0.0035	-0.0035	-0.0035
$7\pi/5$		-0.0001	-0.0017	-0.0002
	-0.0009	-0.0009	-0.0009	-0.0009
$8\pi/5$		+0.0020	+0.0021	+0.0022
	+0.0034	+0.0033	+0.0033	+0.0033
$9\pi/5$		0.0066	0.0201	0.0076
	0.0132	0.0138	0.0136	0.0137
2π		0.0781	0.1092	0.1381
		0.3652	0.5124	0.6505

(b) $c = 3$

Table 1. Comparison of direct sums of series (upper entries) with sums obtained by applying formula (16) ($\phi \neq 0, 2\pi$) and formula (17) ($\phi = 0, 2\pi$) (lower entries).

of the sum of the series, but for $\phi \neq 0$ the use of the transformation (16) is effective. At $\phi = 0$ (or 2π) this formula is singular and

N	S_N	$Sh(S_N)$	$Sh^{(2)}(S_N)$	Sh(Diag.)
10	0.255666			
11	0.259845	0.274032		
12	0.263073	0.274103	0.274250	0.274253
13	0.265570	0.274151	0.274254	
14	0.267504	0.274184	0.274256	0.274257
15	0.269004	0.274206	0.274257	
16	0.270168	0.274222	0.274258	0.274258
17	0.271073	0.274233	0.274258	
18	0.271776	0.274240	0.274258	0.274259
19	0.272323	0.274246		
20	0.272749			

(a) c = 1

N	S_N	$Sh(S_N)$	$Sh^{(2)}(S_N)$	(Sh(Diag.)	$Sh(S_N, Sh(S_N), Sh(Diag))$
43	0.828933				
44	0.839618	1.3625			
45	0.850090	1.3650	1.4103	1.4150	1.4204
46	0.860353	1.3673	1.4107	1.4150	1.4200
47	0.870412	1.3695	1.4113	1.4154	1.4201
48	0.880273	1.3716	1.4118	1.4156	1.4199
49	0.889940	1.3735	1.4121	1.4157	1.4197
50	0.899417	1.3754	1.4128	1.4162	1.4201
51	0.908709	1.3772	1.4129	1.4161	1.4196
52	0.917820	1.3789			
53	0.926755				

(b) c = 2

Table 2. Application of Shanks transformations in various ways to series in (14).

cannot be used. Hence at these points, where the series is monotonic, we apply a Shanks transformation,

$$Sh(S_N) = \frac{S_{N+1}\, S_{N-1} - S_N^2}{S_{N+1} + S_{N-1} - 2S_N}, \tag{17}$$

to the partial sums S_N and the value so obtained is also entered in Tables 1 and 2 at $\phi = 0$ and 2π. Although, at $c = 1$, this transformation provides significantly improved convergence it is clear that at $c = 2$ and $c = 3$ more needs to be done. Hence in Table 2. the effect of applying Shanks transformations in various ways is summarized. A second Shanks iteration $Sh(Sh(S_N)) = Sh^{(2)}(S_N)$ is given in column 4 and one on the 'diagonal' elements of columns 2, 3 and 4 given in column 5 (see [7]). Finally, in Table 2, the effect of the same transformation on columns 2,3 and 5 is given in column 6. There is a marked improvement in convergence when $c = 1$ and when $c = 2$ a strong indication that the sum is 1.419 or 1.420, although the results are not sufficiently accurate to be certain which. Corresponding results for $c = 3$ are not yet complete.

7. DISCUSSION

We first observe that the role of $\varepsilon(c)$, representing the dimensionless pressure averaged over a cycle, is a rather complicated one in the analysis. As Table 3 indicates it increases with c above its value 1 for incompressible flow and hence its position in both the outer layer equations (9) and (11) and inner layer solution (13) suggest that main stream compressibility leads to a thickening of each of these layers.

c	0.5	1	2	3
ε	1.04469	1.17926	1.72530	2.66344

Table 3.

The leading order heat transfer is governed by Q and the general behaviour of Q over a cycle, a positive maximum and negative minimum of smaller magnitude, holds for all values of c. However, the maximum increases enormously at the larger values of c until at $c = 3$ Q is very 'peaky' having a large maximum and very small negative values over much of the cycle. The actual leading order heat transfer includes a factor $(1-2c \cos \bar{t}/7)^{9/2}$ as well and this tends to enhance the negative parts of Q, which occur roughly over the middle range of the cycle, relative to the positive parts.

A result of practical importance which emerges is the appearance of a contribution of $O(s^{\frac{1}{2}})$ to the mean value of q_w over a cycle. From [4] it arises because of the interaction of variations in fluid properties and wall temperature gradient and the effect increases considerably with c.

The leading order terms are independent of the form of $U_e(x)$; however it is evident from equations (9) and (11) that the determination of higher order terms would require the form of U_e to be specified.

Finally, the analysis can be extended to include the calculation of the skin friction.

REFERENCES

1. C.R. Illingworth; The effects of a sound wave on the compressible boundary layer on a flat plate. J. Fluid Mech. 3, 471-93 (1958).
2. P.D. Richardson; Effects of sound and vibrations on heat transfer. App. Mechs. Revs. 20, 201-17 (1967).
3. D.P. Telionis and T.R. Gupta; Compressible oscillating boundary layers. AIAA Jour. 15, 974-83 (1977).
4. R.J. Gribben; Finite amplitude high frequency oscillations in a compressible oscillating boundary layer. Q.J.M.A.M. XXXII, 283-302 (1979).
5. R.J. Gribben; On oscillations in a compressible oscillating boundary layer. To be published in Proc. Roy. Soc.Edinburgh.
6. J.E. Kiefer and G.H. Weiss; A comparison of two methods for accelerating the convergence of Fourier series. (Private communication).
7. M. Van Dyke; Analysis and improvement of perturbation series. Q.J.M.A.M. XXVII, 4, 423-450 (1974).

Dr. R.J. Gribben,
Department of Mathematics,
University of Strathclyde,
Glasgow. Scotland.

Numerical solution of scattering problems by integral equation methods

Z. REUT

SUMMARY

The Helmholtz integral equation formulation of the exterior acoustic problem with the hard boundary condition is reviewed. Two methods for the numerical determination of boundary values are dealt with: the well-known CHIEF method, and the newly implemented CONDOR method. The latter is basically the method of Burton and Miller (1971) incorporating the results of Terai (1980), and it combines the surface Helmholtz equation with its normal derivative form. Although both methods fail for plate-like structure surfaces very thin compared with wavelength, the CONDOR method can be applied to thinner plates than the CHIEF method. While there are serious problems with the CHIEF method concerning the selection of interior collocation points as long as the surface mesh is sufficiently fine.

1. INTRODUCTION

Numerical methods for the exterior acoustic problem derive their importance from the fact that rigorous analytical solutions can only be found for structure surfaces of simple shapes. Moreover, the analytical solutions are often given in the form of finite series of slow convergence and contain special functions that are not suitable for straightforward calculations. The problem is particularly acute when the size of the structure surface is comparable to the wavelength of the incident field.

Integral equation (boundary element) method has some advantages over finite element as well as finite difference method in the case of exterior problems (Burton 1976). The main problem is the non-uniqueness of the exterior solution for wavenumbers coinciding with the characteristic wavenumbers of the corresponding interior problem. However, there are a number of ways to remedy this difficulty in the usual Helmholtz representation.

In the CHIEF (Combined Helmholtz Integral Equation Formulation) method (Schenck 1968) the surface Helmholtz equation (SHE) is combined with the interior Helmholtz relation (IHR) collocated at a finite number of suitably selected interior points. But Jones (1974), as well as some other investigators, pointed out that the selection of these interior points is a tricky one since no rule has been given on how many should be chosen and where they should be located.

Thus it seems that it would be advantageous to choose among the methods discussed by Burton (1973) that ensure the uniqueness of solution for all wavenumbers. But all these methods involve integral equations with highly singular kernels (due to the second derivative of Green's function) which make numerical implementations difficult. The situation is now different due to a result of Terai (1980) which enables the numerical evaluation of a singular integral of that type. The SHE can linearly be combined with its normal derivative form (NDSHE) according to Burton and Miller (1971), and the optimum value of the coupling coefficient determined. This method, dubbed CONDOR (Composite Outward Normal Derivative Relation) by us, is suitable for numerical implementations and appears to be superior to the CHIEF method.

2. INTEGRAL EQUATION FORMULATION

Let us consider the time harmonic ($e^{i\omega t}$) propagation of small amplitude acoustic waves in an ideal three-dimensional homogeneous medium exterior to a smooth, bounded, closed surface S. Let $p_{in}(\underset{\sim}{x})$ be pressure due to a given incident acoustic wave, and $p_{sc}(\underset{\sim}{x})$ pressure due to the scattered acoustic wave produced by S. Then the total acoustic pressure is given by

$$p(\underset{\sim}{x}) = p_{in}(\underset{\sim}{x}) + p_{sc}(\underset{\sim}{x}), \tag{2.1}$$

and satisfies the Helmholtz integral formula as well as its differentiated from (Burton 1973, Terai 1980).

Specializing to an acoustically rigid structure surface S, the hard boundary condition has to be assumed, i.e.

$$\frac{\partial}{\partial n} p(\underset{\sim}{x}) = 0, \quad \underset{\sim}{x} \in S, \tag{2.2}$$

where $\underset{\sim}{n}$ is the outward normal unit vector. In this special case the Helmholtz integral formula and its differentiated form reduce to

$$p_{in}(\underset{\sim}{x}) + \int_S p(\underset{\sim}{x}_q) \frac{\partial}{\partial n_q} G_k(\underset{\sim}{x},\underset{\sim}{x}_q)\, dS_q = \begin{Bmatrix} p(\underset{\sim}{x}), & \underset{\sim}{x} \in E \\ \frac{1}{2}p(\underset{\sim}{x}), & \underset{\sim}{x} \in S \\ 0, & \underset{\sim}{x} \in I \end{Bmatrix} \tag{2.3}$$

$$\frac{\partial}{\partial n} p_{in}(\underset{\sim}{x}) + \int_S p(\underset{\sim}{x}_q) \frac{\partial^2}{\partial n \partial n_q} G_k(\underset{\sim}{x},\underset{\sim}{x}_q)\, dS_q = \begin{Bmatrix} \frac{\partial}{\partial n} p(\underset{\sim}{x}), & \underset{\sim}{x} \in E \\ 0, & \underset{\sim}{x} \in S \\ 0, & \underset{\sim}{x} \in I \end{Bmatrix} \tag{2.4}$$

respectively. Here E denotes the exterior, and I the interior of the surface S. The unit normal n_q is pointed outward (from S into E) at the point x_q on S. The Green's function is given by

$$G_k(\underset{\sim}{x},\underset{\sim}{x}_q) = \frac{e^{-ik|\underset{\sim}{x} - \underset{\sim}{x}_q|}}{4\pi|\underset{\sim}{x} - \underset{\sim}{x}_q|}, \tag{2.5}$$

where the wavenumber $k = 2\pi/\lambda$, and λ is the wavelength.

Let us now introduce a mesh subdividing the surface S into m surface elements S_i (i=1,2,...,m) in a regular fashion. Assuming that the pressure $p(\underset{\sim}{x})$ is constant and equal to p_i on the surface element S_i, the total acoustic pressure can be approximated by

$$p(x) = \sum_{i=1}^{m} p_i \psi_i(\underset{\sim}{x}), \quad \underset{\sim}{x} \in S, \tag{2.6}$$

where

$$\psi_i(x) = \begin{Bmatrix} 1, & \underset{\sim}{x} \in S_i \\ 0, & \underset{\sim}{x} \notin S_i \end{Bmatrix}. \tag{2.7}$$

3. BOUNDARY VALUES DETERMINATION

Let us assume that $p_{in}(\underset{\sim}{x})$ is a given function of position $\underset{\sim}{x}$. Then any exterior problem can be solved by using the Helmholtz formula providing $p(x)$ and $\frac{\partial}{\partial n} p(\underset{\sim}{x})$ are known on S. In the case of the hard boundary condition, it is sufficient to know $p(\underset{\sim}{x})$ on S only.

But the determination of boundary values is generally a formidable task, difficulties arising due to (i) the non-uniqueness of solutions and (ii) the singularity of kernels. It is well-known that SHE (NDSHE) has no unique solution for wavenumbers coinciding with the characteristic wavenumbers k'(k") of the interior Dirichlet (Neumann) problem (Burton 1973). The kernel of equation (2.3) is mildly singular, while that of equation (2.4) is highly singular for $\underset{\sim}{x} \varepsilon S$. We are describing two numerical methods for determining the boundary values and specializing to the case of the hard boundary condition.

The CHIEF method. The collocation of SHE at m surface points (chosen as the centroids of m surface elements) and the collocation of IHR at m' interior points leads to an overdetermined system of linear equations

$$\sum_{j=1}^{m} A_{ij}\, p_j = (p_{in})_i, \tag{3.1}$$

where i=1,2,...,m+m', and the matrix elements are given by

$$A_{ij} = \tfrac{1}{2}\delta_{ij} - \int_{S_j} \frac{\partial}{\partial n_q} G_k(\underset{\sim}{x}_i, \underset{\sim}{x}_q)\, dS_q, \tag{3.2}$$

for $\underset{\sim}{x}_i \varepsilon S_i$ (i=1,2,...,m), $\underset{\sim}{x}_i \varepsilon I$ (i=m+1, m+2,..., m+m'). The overdetermined system (3.1) is solved by the least squares method giving the total acoustic pressure at the surface collocation points.

The problem is that whenever the collocation point $\underset{\sim}{x}_i$ lies within the region of integration S_j (i.e. for i=j) a singular integral arises. However, since all these singularities are mild ones, they can be dealt with numerically in the usual ways (Burton 1976). Jones (1980) has shown that for a flat rectangular surface element S_j the principal value of the jth singular integral vanishes, i.e.

$$\int_{S_j} \frac{\partial}{\partial n_q} G_k(\underset{\sim}{x}_j, \underset{\sim}{x}_q)\, dS_q = 0, \tag{3.3}$$

and the corresponding diagonal element $A_{jj} = 1/2$.

The CONDOR method. The two integral equations, namely SHE and NDSHE, for any particular problem have only one solution in common even when k coincides with both k' and k" (Burton and Miller 1971). The simplest way to form the composite Helmholtz equation is by taking a linear combination of equations (2.3) and (2.4) for $\underset{\sim}{x} \in S$, i.e.

$$\tfrac{1}{2}p(\underset{\sim}{x}) - \int_S p(\underset{\sim}{x}_q)\left[\frac{\partial}{\partial n_q} G_k(\underset{\sim}{x},\underset{\sim}{x}_q) + \alpha \frac{\partial^2}{\partial n \partial n_q} G_k(\underset{\sim}{x},\underset{\sim}{x}_q)\right] dS_q$$

$$= p_{in}(\underset{\sim}{x}) + \alpha \frac{\partial}{\partial n} p_{in}(\underset{\sim}{x}), \tag{3.4}$$

where α is a complex coefficient which may depend on the wave-number k only. It can be proved that the solution of equation (3.4) is unique when $\mathrm{Im}(\alpha) \neq 0$ for real k (Burton 1973, 1976).

The collocation of equation (3.4) at m surface points (the centroids of m surface elements) leads to an ordinary system of linear equations

$$\sum_{j=1}^{m} A_{ij}\, p_j = (p_{in} + \alpha \frac{\partial}{\partial n} p_{in})_i, \tag{3.5}$$

where i=1,2,...,m, and the matrix elements are given by

$$A_{ij} = \tfrac{1}{2}\delta_{ij} - \int_{S_j} \left[\frac{\partial}{\partial n_q} G_k(\underset{\sim}{x}_i,\underset{\sim}{x}_q) + \alpha \frac{\partial^2}{\partial n_i \partial n_q} G_k(\underset{\sim}{x}_i,\underset{\sim}{x}_q)\right] dS_q \tag{3.6}$$

for $\underset{\sim}{x}_i \in S_i$, i,j=1,2,...,m. The system (3.5) is solved algebraically yielding the total acoustic pressure at surface collocation points.

The problem is that whenever the collocation point $\underset{\sim}{x}_i$ lies within the region of integration S_j, a highly singular integral arises. The method of regularization can be used, but it becomes very cumbersome numerically in three-dimensional cases (Burton 1976). Terai (1980) has demonstrated that for a flat rectangular surface element S_i

$$\int_{S_i} \frac{\partial^2}{\partial n_i \partial n_q} G_k(\underset{\sim}{x}_i,\underset{\sim}{x}_q)\, dS_q = -\tfrac{1}{2}ik - \frac{1}{4\pi}\int_0^{2\pi} \frac{e^{-ik\rho_i}}{\rho_i}\, d\theta_i, \tag{3.7}$$

where $\rho_i = \rho_i(\theta_i)$ is the polar equation of the contour C_i of the surface element S_i with respect to its centroid as an origin. In this case the corresponding diagonal matrix element reduces to

$$A_{ii} = \frac{1}{2} + \alpha\left(\frac{1}{2}ik + \frac{1}{4\pi}\int_0^{2\pi} \frac{e^{-ik\rho_i}}{\rho_i}\, d\theta_i\right). \tag{3.8}$$

Let us denote by $\bar{\rho}$ the mean value of ρ_i (averaged over the polar angle θ_i and then over the subscript i). The quantity $k\bar{\rho}$ can be used as a measure of the surface mesh fineness. Some considerations suggest that the optimum value of the coupling coefficient α is $-i/k$.

4. NEAR- AND FAR-FIELD

The scattered pressure at any point in the medium exterior to the surface S is determined by the total pressure and its normal derivative on S according to the Helmholtz integral. In the case of the hard boundary condition only the total pressure on S has to be known and equation (2.3) for $\underset{\sim}{x} \varepsilon E$ should be used. At a near-field point the integral has to be evaluated directly. For a far-field point the well-known asymptotic expression may be used providing the origin of coordinates is taken within the surface S.

Adopting again the piecewise constant approximation to $p(\underset{\sim}{x})$ given by equations (2.6) and (2.7), the near- and far-field approximations for the scattered pressure become

$$p_{sc}(x) = \sum_{i=1}^{m} p_i \int_{S_i} \frac{\partial}{\partial n_q} G_k(\underset{\sim}{x},\underset{\sim}{x}_q)\, dS_q, \tag{4.1}$$

$$p_{sc}(\underset{\sim}{x}) = \frac{e^{-ik|\underset{\sim}{x}|}}{4\pi|x|} \sum_{i=1}^{m} p_i \int_{S_i} ik\underset{\sim}{n}_q \cdot \frac{\underset{\sim}{x}-\underset{\sim}{x}_q}{|\underset{\sim}{x}-\underset{\sim}{x}_q|}\, e^{ik\underset{\sim}{x}\cdot\underset{\sim}{x}_q/|\underset{\sim}{x}|} dS_q, \tag{4.2}$$

respectively, where p_i (i=1,2,...,m) is the total pressure at the ith surface collocation point. Knowing the scattered pressure at $\underset{\sim}{x}$, the total pressure at the same position can immediately be found by equation (2.1).

5. A SIMPLE POINT SOURCE

As an example, let us consider the case of a simple point source at $\underset{\sim}{x}_o$ outside a hard boundary surface S. The incident pressure is now given by

$$p_{in}(\underset{\sim}{x}) = \frac{e^{-ik|\underset{\sim}{x} - \underset{\sim}{x}_o|}}{4\pi|\underset{\sim}{x} - \underset{\sim}{x}_o|}, \tag{4.3}$$

and its derivative in the direction of a unit vector $\underset{\sim}{n}$ reads

$$\frac{\partial}{\partial n} p_{in}(\underset{\sim}{x}) = - \frac{1+ik|\underset{\sim}{x}-\underset{\sim}{x}_o|}{4\pi|\underset{\sim}{x}-\underset{\sim}{x}_o|^3}(\underset{\sim}{x}-\underset{\sim}{x}_o).\underset{\sim}{n}\ e^{-ik|\underset{\sim}{x}-\underset{\sim}{x}_o|}. \tag{4.4}$$

Substituting equation (4.3) into equation (3.1) when using the CHIEF method, or equations (4.3) and (4.4) into equation (3.5) when using the CONDOR method, it is clear that the exterior problem can be solved numerically as outlined in Sections 3 and 4.

6. NUMERICAL IMPLEMENTATION

Both methods described in Section 3 have been implemented, tested, and applied to some scattering problems, and the following is an account of our experience at the University of Dundee.

The CHIEF method was implemented and tested by Wilton (1975). Goddard (1977) noted that the singular integrals appearing in equation (3.2) have very small values compared with the non-singular integrals and thus could probably be neglected in most practical cases. This is in agreement with equation (3.3)

The CONDOR method was implemented and tested by Vrcelj (1981) along the lines of Wilton (1975) and Goddard (1977). Numerical tests have shown that for a given surface mesh the best accuracy solutions are obtained with $\alpha = -i/k$. Our experience tells us that $k\bar{\rho} \lesssim 1$ has to be satisfied to obtain solutions of engineering accuracy.

We have found that both methods fail for plate-like structure surfaces very thin compared with wavelength, but that the CONDOR method can be applied to thinner plates than the CHIEF method. The CHIEF (CONDOR) method in our implementation requires $ka>0.5$ (0.1), where a is the half-thickness of the plate. We have encountered serious problems with the CHIEF method concerning the selection of interior collocation points for wavenumbers greater than k'_{min} (the lowest characteristic wavenumber of the interior Dirichlet problem). No problems appear with the CONDOR method as long as the surface mesh is sufficiently fine, i.e. $k\bar{\rho} \lesssim 1$.

Thus it seems that the CONDOR method is superior to the CHIEF method to the point when the latter should be abandoned. The CONDOR method is the only one we are presently aware of that can bridge the gap between very low frequencies when nearly any method based on the classical wave theory is usable and very high frequencies when the ray theory method becomes advantageous.

ACKNOWLEDGEMENT

This research has been supported by the Ministry of Defence.

REFERENCES

1. Burton, A.J. (1973) NPL Report NAC30.
2. Burton, A.J. (1976) NPL Contract: OC5/535.
3. Burton, A.J. and Miller, G.F. (1971) Proc. Roy. Soc. London A323, 201-210.
4. Goddard, A.J. (1977) MOD Res. Report 2, University of Dundee.
5. Jones, D.S. (1974) Q. Jl. Mech. Appl. Math. 27, 129-142.
6. Jones, D.S. (1980) Private Communication.
7. Schenck, H.A. (1968) J. Acoust. Soc. Am. 44, 41-58.
8. Terai, T. (1980) J. Sound Vib. 69(1), 71-100.
9. Vrcelj, Z. (1981) MOD Res. Report 2, University of Dundee.
10. Wilton, D.T. (1975) MOD Res. Report 6, University of Dundee.

*Dr. Z. Reut,
Department of Mathematics,
University of Dundee,
Dundee. Scotland.

*School of Physics, University of Bath, Bath BA2 7AY from 1st October, 1981.

Acoustic fluid flow through a wall perforated by small holes

J. SANCHEZ-HUBERT

We consider the unsteady, incompressible, slightly viscous fluid flow in a domain Ω containing a wall with many small holes. The distance between two contiguous holes is of order η and the size of each hole is of order ε. η and ε are two small parameters.

ACOUSTIC FLOW THROUGH A PERFORATED WALL IN THE THREE-DIMENSIONAL CASE

Let us consider a domain Ω of R^3 bounded with smooth boundary. The coordinates are x_i (i = 1, 2, 3), the origin is contained in Ω.

For the time being we suppose that

$$\varepsilon, \eta \to 0 , \qquad \frac{\varepsilon}{\eta} \to 0 \tag{1}$$

In the plane $x_3 = 0$, which is the wall, we consider a periodic array of points of periods ηa and ηb.

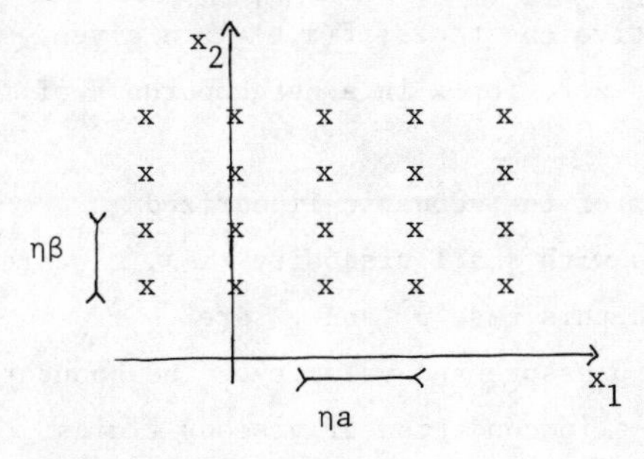

Then we have :

$$\begin{cases} x_1 = na\eta \\ x_2 = mb\eta \end{cases} \tag{2}$$

where n and m are integers. Now we consider a system of holes εS centred at the points defined by (2). In the plane $x_3 = 0$ we thus have holes, of size ε, periodically located at distances of order η. From the last relation (1) the distance between two adjacent points is large with respect to the size of the hole.

Before continuing we make the following remark: We assume the wall to be the plane $x_3 = 0$, more generally a curved wall can be considered. Indeed, in the ε-dilatation any small portion of the surface is equivalent to its tangent plane.

Now we define the domain Ω^ε, it is obtained by removing from Ω the perforated wall. This domain depends on ε (and on η but we shall see that η will be considered as a function of ε).

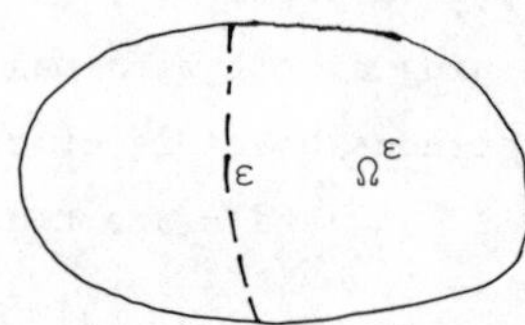

We search for the functions v^ε and p^ε defined for $x \in \Omega^\varepsilon$, $t \in (0,\infty)$ satisfying the boundary value problem (3)-(4).

$$\begin{cases} \rho \dfrac{\partial \underline{v}}{\partial t} = - \underline{\text{grad}}\; p^\varepsilon + \varepsilon^2 \lambda \Delta \underline{v}^\varepsilon + \varepsilon^2 \mu\; \underline{\text{grad}}\; \text{div}\; v^\varepsilon + \underline{f} \\ \dfrac{\partial p^\varepsilon}{\partial t} + \rho c^2 \text{div}\; \underline{v}^\varepsilon = 0 \end{cases} \tag{3}$$

$$\begin{cases} \underline{v}^\varepsilon = 0 \quad \text{on} \quad \partial\Omega^\varepsilon \\ \underline{v}^\varepsilon,\; p^\varepsilon = 0 \quad \text{for } t = 0 \end{cases} \tag{4}$$

ρ, λ, μ and c^2 are given positive constants; $\underline{f}(x,t)$ is a given function, we assume that it is zero for x in a neighbourhood of the wall $x_3 = 0$ and any t.

The problem (3)-(4) is that of the acoustic linearized vibrations of a barotropic gas with small viscosity ($\varepsilon^2\lambda$, $\varepsilon^2\mu$ are the viscous coefficients). In this case p^ε and $\underline{v}^\varepsilon$ are respectively perturbations of pressure and velocity. The boundary condition is the classical no-slip condition of viscous fluids. Finally, instead of $f = 0$ in the vicinity of the wall, we may consider non zero initial or boundary conditions.

<u>As $\varepsilon \to 0$</u> a boundary layer appears in the vicinity of $\partial\Omega^\varepsilon$, out of this region we consider an expansion (<u>outer expansion</u>) of the form (5).

$$\begin{cases} \underline{v}^{\varepsilon} = \underline{v}^{o}(x,t) + \varepsilon \underline{v}^{1}(x,t)+\dots , \\ p^{\varepsilon} = p^{o}(x,t) + \varepsilon p^{1}(x,t)+\dots . \end{cases} \quad (5)$$

Then (3) and (4) give, to order ε^{o} :

$$\begin{cases} \rho \dfrac{\partial \underline{v}^{o}}{\partial t} = - \text{grad } p^{o} + \underline{f} \\ \dfrac{\partial p^{o}}{\partial t} = \rho c^{2} \text{div } \underline{v}^{o} = 0 \end{cases} \quad (6)$$

with

$$\underline{v}^{o}(x,0), \quad p^{o}(x,0) = 0 \quad (7)$$

The appropriate boundary conditions will be obtained by matching with the boundary layer solution.

The boundary of Ω^{ε} consists of the standard wall $\partial\Omega$ enclosing Ω and the region $x_3 = 0$ formed by the perforated wall. In the first one a classical oscillating boundary layer appears and we have

$$\underline{v}^{o} \cdot \underline{n} = 0 \quad \text{on} \quad \partial\Omega \quad (8)$$

On the perforated wall, we assume:

$$\begin{cases} [\![p^{o}]\!] \neq 0 \\ v_3^{o+} = v_3^{o-} \neq 0 \end{cases} \quad \text{on } x_3 = 0 \text{ (Perforated wall)} \quad (9)$$

where

$$\begin{cases} [\![p^{o}]\!] = p^{o+}(x_1,x_2,t) - p^{o-}(x_1,x_2,t) \\ p^{o\pm} (\text{resp. } v_3^{o\pm}) = \lim\limits_{x_3 \to 0^{\pm}} p^{o}(x,t) (\text{resp. } v_3^{o}(x,t)) \end{cases}$$

The meaning of (9) is the following: If the holes are asymptotically small the wall behaves in the limit as an impermeable one and we have $[\![p^{o}]\!] \neq 0$, $v_3^{o+} = v_3^{o-} = 0$. On the contrary if the holes are asymptotically large the effect of the wall disappears in the limit and we have $[\![p^{o}]\!] = 0$, $v_3^{o+} = v_3^{o-} \neq 0$. Consequently (9) are intermediate boundary conditions.

<u>In the vicinity of $x_3 = 0$(η-layer)</u> we consider a new variable $y = x/\eta$ and as in the homogenization method, we search for p^{ε} and v^{ε} two-scale expansions of the form (11).

$$\begin{cases} p^{\varepsilon} = \hat{p}^{o}(x,y,t) + \eta \hat{p}^{1}(x,y,t)+\dots , \\ \underline{v}^{\varepsilon} = \hat{v}^{o}(x,y,t) + \eta \hat{\underline{v}}^{1}(x,y,t)+\dots . \end{cases} \quad (11)$$

where the functions depend on $x = (x_1,x_2)$ and on $y = (y_1,y_2,y_3)$; $y_i = x_i/\eta \quad i = 1,2,3.$

To the first order, from (3)-(4) we obtain

$$\begin{cases} \mathrm{grad}_y\ \hat{p}^o = 0 \qquad (\Rightarrow \hat{p}^o = \hat{p}^o(x_1,x_2,t)) \\ \rho \dfrac{\partial \underline{\hat{v}}^o}{\partial t} = - \underline{\mathrm{grad}}_x\ \hat{p}^o - \underline{\mathrm{grad}}_y\ \hat{p}^1 \\ \mathrm{div}_y\ \underline{\hat{v}}^o = 0 \end{cases} \tag{12}$$

which shows that the flow is an incompressible and inviscid one at this order.

Now if we consider x as a parameter the second equation (12) may be written

$$\rho \frac{\partial \underline{\hat{v}}^o}{\partial t} = - \underline{\mathrm{grad}}_y\ (\hat{p}^1 + \underline{y}.\ \underline{\mathrm{grad}}_x\ \hat{p}^o) \tag{13}$$

This equation and the last equation (12) show that the flow in this region (η-layer) is that of an incompressible fluid with sources at the points

$$y_1 = a\,n\ , \quad y_2 = b\,m\ , \quad y_3 = 0$$

and tending to a uniform v^{o+} flow as $y_3 \to \infty$. Such a flow was studied by Tuck (with $v_o^+ = 0$); he gives the asymptotic behaviour of the velocity potential as $y_3 \to \infty$. We thus have

$$\hat{p}_1 = \gamma\phi(y); \qquad \phi \sim \frac{y_3 + \mathrm{cost}}{2ab} \quad \text{as} \quad y_3 \to \infty \tag{14}$$

Then (14) and $(12)_2$ give (15) and the matching with the outer expansion gives finally (16).

$$\begin{cases} \lim_{y_3 \to +\infty} \ \rho \dfrac{\partial \hat{v}_i^o}{\partial t} = - \dfrac{\partial p^{o+}}{\partial x_i} \qquad i = 1,2 \\ \lim_{y_3 \to +\infty} \ \rho \dfrac{\partial \hat{v}_3^o}{\partial t} = - \dfrac{\gamma}{ab} \end{cases} \tag{15}$$

$$\begin{cases} \lim_{y_3 \to +\infty} \ \hat{v}_3^o = v_3^{o+} \quad \Leftarrow \quad (15)_1 \\ \hat{p}^o(x,t) = p^{o+}(x_1,x_2,t) \quad \Leftarrow \quad (12)_1 \\ \hat{p}^1(x,y,t) = -2ab\rho \dfrac{\partial v_3^{o+}}{\partial t}\ \phi(y) + c(x,t) \quad \Leftarrow \quad (13) \end{cases} \tag{16}$$

Consequently <u>in the η-layer we have</u> (17)

$$\boxed{p^\varepsilon = p^{o+} - 2ab\ \rho \frac{\partial v_3^{o+}}{\partial t}\ \eta\phi(y) + \theta(\eta); \quad \theta(\eta) << \eta} \tag{17}$$

An analogous study holds for $x_3 < 0$.

Now we take the origin at the centre of a hole and we introduce the local (or inner) variable $z = x/\varepsilon$. We then search for $\underline{v}^\varepsilon$ and p^ε expansions of the form

$$\begin{cases} \underline{v}^\varepsilon = \underline{V}(x_1,x_2,z,t)+\ldots , \\ p^\varepsilon = P(x_1,x_2,z,t)+\ldots \end{cases} \qquad z = \frac{x}{\varepsilon} \tag{18}$$

We remark that in (18) the functions do not depend on y; we shall see later that it is a natural consequence of the matching conditions.

From the second equation (3) we obtain

$$\operatorname{div}_z \underline{V} = 0 \qquad (19)_1$$

which shows that, to the first order, the flow in the hole is incompressible (as well as in the η-layer). Consequently the flux of $\underline{v}^\varepsilon$ through each hole is the same as the flux through the η-layer, we thus have

$$\varepsilon^2 \int_S V_3 \, dz_1 \, dz_2 = \eta^2 \, ab \, v_3^{o+} \qquad (19)_2$$

and $\underline{V}$ must be of order $(\eta|\varepsilon)^2$.

We remark that the assumption $v_3^{o+} = v_3^{o-}$ is here justified.

On the other hand we have (see Tuck)

$$\eta\phi(y) \sim \eta \frac{1}{2\pi|y|} \qquad \text{for } |y| \to 0$$

therefore we can write

$$\eta\phi(y) \sim \frac{\eta^2}{\varepsilon} \frac{1}{2\pi|z|} \tag{20}$$

Now if we come back to the first equation (3):

$$\left(\frac{\eta}{\varepsilon}\right)^2 \frac{\partial \underline{V}}{\partial t} = -\frac{1}{\varepsilon} \underline{\operatorname{grad}}_z P + \lambda \frac{\eta^2}{\varepsilon^2} \Delta_z \underline{V}$$

We see that the critical behaviour appears for $\eta^2 = O(\varepsilon)$; we shall take

$$\boxed{\eta^2 = \varepsilon}$$

Then, the matching of p^ε in (18) with p^ε in (17), taking account of (20), shows that:

$P(x_1,x_2,z,t) \to p^{o\pm}(x_1,x_2,t)$ for $|z_3| \to \infty$

we can consider that the η-layer "vanishes" to the first order, this is the reason for taking $\underline{V}$ and P independent of y.

Now we consider the <u>local problem</u> (21)-(26).

$$\rho \frac{\partial \underline{V}}{\partial t} = \underline{\text{grad}}_z P + \lambda \Delta_z \underline{V} \tag{21}$$

$$\text{div}_z \underline{V} = 0 \tag{22}$$

$$\underline{V} \to \underline{0} \quad \text{for} \quad |z| \to \infty \tag{23}$$

$$P(x_1,x_2,z,t) \to p^{o\pm}(x_1,x_2,t) \quad \text{for} \quad \begin{cases} |z| \to \infty \\ z_3 \to \pm\infty \end{cases} \tag{24}$$

$$\begin{cases} \underline{V} = \underline{0} \quad \text{on} \quad \partial \mathbb{R}^3_s \quad \text{for } \lambda > 0 \\ V_3 = 0 \quad \text{on} \quad \partial \mathbb{R}^3_s \quad \text{for } \lambda = 0 \end{cases} \tag{25}$$

$$\underline{V}(z,0) = 0 \tag{26}$$

We define the <u>function spaces</u> H and V:

$$H = \{\underline{u} \in \underline{L}^2(\mathbb{R}^3_s); \quad \text{div}\, \underline{u} = 0, \quad u_3|_{\partial \mathbb{R}^3_s} = 0\}$$

$$V = \{\underline{u} \in \underline{H}^1_o(\mathbb{R}^3_s); \quad \text{div}\, \underline{u} = 0\}$$

$$\mathbb{R}^3_s = \{z \in \mathbb{R}^3; \quad \text{either} \quad z_z \neq 0 \quad \text{or} \quad (z_1,z_2) \in S\}$$

which are closed subspaces of $\underline{L}^2$ and $\underline{H}^1_o$ respectively and thus Hilbert spaces for the corresponding scalar products.

To solve the local problem we search for $\underline{V}(t)$ function of t with values in H; (22), (23) and (25) are then automatically fulfilled. Instead of P we introduce a new function Q which is defined by (27).

$$Q = P - [\![p^o]\!]\, q \tag{27}$$

In (27) q is a smooth function defined on $\mathbb{R}^3_s$ such that for sufficiently large $|z|$ it takes the value 1 for $z_3 > 0$ and 0 for $z_3 < 0$. Then under the preceeding hypotheses we prove lemma 1.

<u>Lemma 1</u> : $\displaystyle\int_{\mathbb{R}^3_s} \frac{\partial q}{\partial z_i} V_i \; dz$

is a bounded non zero functional on H. Consequently, this functional is expressible as a scalar product by a well determined non zero element F of H. Moreover the value of this functional is equal to the flux of $\underline{V}$ through the hole i.e.

$$(\underline{F},\underline{V})_H = \int_{\mathbb{R}^3_s} \frac{\partial q}{\partial z_i} V_i \, dz = \int_s V_3 \, dz_1 \, dz_2 = \phi(\underline{V})$$

According to the standard theory of Navier-Stokes equations the pressure term in (21) will be eliminated by orthogonality. We consider the orthogonal complement $H^{\perp}$ of H in $L^2(R^3_s)$ and we prove that, in some generalized sense, it consists of the gradient of the functions which tend to a constant at infinity, the same constant in the two regions $z_3 > 0$ and $z_3 < 0$. We then have

Lemma 2: Let $H^{\perp}$ be the orthogonal complement of H in $L^2(R^3_s)$ there exists a function $Q \in H^1_{loc}$ defined up to additive constant such that

$\underline{V} = \underline{\text{grad}}\ Q$; $Q \to C$ for $|z| \to +\infty$

Then the condition (24) and the equation (21) respectively becomes

$$Q \to p^{o-} \quad \text{for } |z| \to +\infty \tag{28}$$

$$\rho \frac{\partial \underline{V}}{\partial t} - \lambda\Delta\, \underline{V} = - \underline{\text{grad}}\ Q - [\![p^o]\!]\ \underline{\text{grad}}\ q \tag{29}$$

If we multiply (29) by a test function $\underline{w} \in V$, by integrating by parts we have $(30)_1$ where F is the element of H above defined. Conversely if $\underline{V}$ satisfies $(30)_1$ for any $\underline{w} \in V$ and $\Delta\underline{V} \in L^2$ we see, by Lemma 2, that Q which satisfies (27) exists and is well determined. In the Lemma 2 Q is determined up to additive constant but this constant may be chosen for each t in order to satisfy (27).

We then are in the standard situation of parabolic equations: If A is the self-adjoint operator of H associated with the form $a(\underline{V},\underline{w})$ $(30)_1$ amounts to $(30)_2$:

$$\rho(\frac{\partial \underline{V}}{\partial t}, \underline{w})_H + \lambda a(\underline{V}, \underline{w}) = - [\![p^o]\!]\ (\underline{F}, \underline{w})_H \quad \forall \underline{w} \in V \tag{30_1}$$

$$\rho \frac{d\underline{V}}{dt} + \lambda A\underline{V} = - [\![p^o]\!] (t)\ F \tag{30_2}$$

We solve $(30)_2$ with the initial condition (26) by standard semigroup theory from which

$$\underline{V}(t) = \frac{1}{\rho} \int_o^t e^{-\frac{\lambda}{\rho} A(t-s)} [\![p^o]\!]\ (s)\ \underline{F}\, ds \tag{31}$$

We have then proved that the local problem (21)-(26) has a unique solution given by (31). We remark that the boundary conditions

$[\![p^o]\!]$ (s) for $s \in (0,t)$ and the initial values $\underline{V}(0)$ determine $\underline{V}(t)$. The flux of $\underline{V}$ through the hole is defined by

$$\phi(\underline{V}(t)) = (\underline{F},\underline{V}) = - \int_o^t g(t-s) \; [\![p^o]\!] (s) \; ds \tag{32}$$

where

$$g(\xi) = \frac{1}{\rho} (e^{-\frac{\lambda}{\rho} A\xi} \underline{F}, \underline{F})_H$$

This also holds in the inviscid case but $\underline{V}$ is then merely the primitive function of $[\![p^o]\!]$ (t) $\underline{F}$.

At last, from $(19)_2$ with $\eta^2 = \varepsilon$ we have $(19)_3$ which gives with $(19)_4$ and (32) the convolution relation between v_3^{o+} and $[\![p^o]\!]$ (33)

$$\int_s V_3 \, dz_1 dz_2 = ab \; v_3^{o+} \tag{19$_3$}$$

$$P(x_1,x_2,z,t) \underset{z_3 \to \pm\infty}{\to} p^{o\pm}(x_1,x_2,t) \tag{19$_4$}$$

$$\phi(\underline{V}) = ab \; v_3^{o+}(x_1,x_2 t) = - \int_o^t g(t-s) \; [\![p^o]\!] (s) \; ds \tag{33}$$

Now we shall soon prove the <u>existence and uniqueness of the solution of the outer problem</u>. We consider the equations (6) with the boundary and initial conditions (7) namely (34) and (35) with $[\![p^o]\!] \neq 0$, $v_3^{o+} = v_3^{o-}$ on $x_3 = 0$.

$$\begin{cases} \rho \dfrac{\partial \underline{v}^o}{\partial t} = - \underline{\text{grad}} \; p^o + \underline{f} \\[2ex] \dfrac{\partial p^o}{\partial t} = \rho \, c^2 \; \text{div} \; \underline{v}^o = 0 \end{cases} \tag{34}$$

$$\begin{cases} \underline{v}_o^+ \cdot \underline{n}\big|_{\partial\Omega^\pm} = 0 \\[1ex] v_3^o = \displaystyle\int_o^t g(t-s) \, [\![p^o]\!] (s) \; ds \\[1ex] p^o(x,o) = 0, \quad \underline{v}^o(x,o) = 0 \end{cases} \tag{35}$$

This system is therefore integro-differential and, in some sense, of hyperbolic type. Instead of studying the existence and uniqueness of the solution under this form, we shall study the system formed by $\underline{v}^o$, p^o and $\underline{V}$ simultaneously.

In order to simplify we shall take $a = b = c^2 = \rho = \lambda = 1$ and we write (36)

$$\begin{cases} \underline{v}^o = \dfrac{\partial \underline{u}^o}{\partial t}, & \underline{u}^o(o) = 0 \\ \underline{V} = \dfrac{\partial \underline{U}}{\partial t}, & \underline{U}(o) = 0 \end{cases} \qquad (36)$$

then, the second equation (33) becomes the first one (37) from which the first equation (34) take the form $(37)_2$.

$$\begin{cases} p^o = -\operatorname{div} \underline{u}^o \\ \rho \dfrac{\partial^2 \underline{u}^o}{\partial t^2} = \underline{\operatorname{grad}} \operatorname{div} \underline{u}^o + \underline{f} \end{cases} \qquad (37)$$

Finally the equation satisfied by $\underline{V}$ can be written in the form

$$\begin{cases} \dfrac{\partial \underline{V}}{\partial t} + A\,\underline{V} = [\![\operatorname{div} \underline{u}^o]\!]\, \underline{F} \\ (F, \underline{U})_H = u_3^o \quad \text{on} \quad x_3 = 0 \end{cases} \qquad (38)$$

Now we define u^1, u^2, u^3, u^4:

$$u^1 = \underline{u}^o, \quad u^2 = \underline{U}, \quad u^3 = \frac{\partial \underline{u}^o}{\partial t}, \quad u^4 = \frac{\partial \underline{U}}{\partial t}$$

and we obtain the system which we shall soon study, namely (39) or (40) with (41).

$$\begin{cases} \dfrac{du^1}{dt} = u^3 \\ \dfrac{du^2}{dt} = u^4 \\ \dfrac{du^3}{dt} = \operatorname{grad} \operatorname{div} u^1 + f \\ \dfrac{du^4}{dt} = Au^4 + [\![\operatorname{div} u^1]\!]\, F \end{cases} \qquad (39)$$

$$\frac{du}{dt} = -\mathcal{A}\, u + f \qquad (40)$$

$$\begin{cases} u^{-1}.n = 0 \quad \text{on} \quad \partial\Omega \\ u_3^1 = (F,u)_H \quad \text{on } \Sigma \end{cases} \qquad (41)$$

We can prove that this system is in the classical framework of semigroup theory.

We define the space $\mathcal{V}$ where $H(\Omega, \operatorname{div})$ is the classical space formed by the vectors of $L^2(\Omega)$ whose divergence belongs to $L^2(\Omega)$.

$$\mathcal{V} = \{\underline{u}^1, \underline{u}^2;\ \underline{u}^1 \in H(\Omega,\text{div}),\ \underline{u}^2 \in L^2(\Sigma,V),$$

$$\underline{u}^1 \cdot \underline{n}|_{\partial\Omega} = 0,\quad (\underline{F},\underline{u}^2)_H = u_3^1 \quad \text{on } \Sigma\}$$

$$\| u^1, u^2 \|^2_{\mathcal{V}} = \int_\Omega (|\underline{u}^1|^2 + |\text{div}\ \underline{u}^1|^2)dx + \int_\Sigma \| u^2 \|^2_V \ dx_1 dx_2 \qquad (42)$$

It is clear that the conditions (41) are included in $\mathcal{V}$. Moreover there are linear relations between traces belonging to $H^{-\frac{1}{2}}(\partial\Omega)$, $L^2(\Sigma)$ and $H^{-\frac{1}{2}}(\Sigma)$ and consequently $\mathcal{V}$ is a Hilbert space for the norm (42). We also define $\mathcal{H}$

$$\mathcal{H} = \text{closure of } \mathcal{V} \text{ in } \underline{L}^2(\Omega) \times L^2(\Sigma, V)$$

$$\| u^1, u^2 \|^2_{\mathcal{H}} = \int_\Omega |u^1|^2 dx + \int_\Sigma \| u^2 \|^2_V \ dx_1 dx_2$$

Now we apply the Lumer-Phillips theorem . under the transformation $u = e^{\alpha t}\ w$ the corresponding system for w involves the operator $\hat{\mathcal{A}} = \mathcal{A} + \alpha I$ instead of $\mathcal{A}$ and we can prove the following theorem.

Theorem: The operator $-\mathcal{A}$ is the generator of a continuous semi-group in $\mathcal{V} \times \mathcal{H}$.

To conclude we shall notice our results concerning the two dimensional case. We have obtained that the critical case is for $\varepsilon = ke^{-\beta/\eta}$ where β and k are positive constants and in this case the matching gives

$$\boxed{[\![p^o]\!] = -\ 2\rho\ \frac{a\beta}{\pi}\ \frac{\partial v_2^{o+}}{\partial t}} \qquad (43)$$

we can notice that in these two cases it is possible to consider a nonperiodic array. Finally in the case $\varepsilon = \eta$ it is evident that the outer flow is the same as if there is no wall nevertheless the preceeding reasonings do not hold because the η and ε regions coincide. This case is in the general framework of homogenization of boundaries and the periodicity on y is essential.

REFERENCES

1. A. Bensoussan, J.L. Lions and G. Papanicolaou: "Asymptotic analysis for periodic structures". North-Holland, Amsterdam (1978).
2. L.D. Landau and E.M. Lifchitz: "Mécanique des fluides". Mir, Moscou (1971).
3. V.A. Marchenko and E. Ya. Khruslov: "Boundary value problems in domains with a granular boundary". Naukova Dumka, Kiev (1974).
4. J. Sanchez-Hubert and E. Sanchez-Palencia: "Accoustic fluid flow through holes and permeability of perforated walls", J. of Math. Anal. and Applic. (to be published).
5. E. Sanchez-Palencia: "Non-homogeneous media and vibration theory", Lect. Notes Phys. 127 Springer-Berlin (1980).
6. R. Temam: "Navier-Stokes equations", North-Holland, Amsterdam.
7. E.O. Tuck: "Unsteady flow of a viscous fluid from a source in a wall". Jour. Fluid Mech. 41 (1970) p. 641-652.
8. E.O. Tuck: "Matching problems involving flow through small holes", in Advan. Appl. Mech. 15, p. 89-158, Academic Press, New York (1975).
9. M. Van Dyke: "Perturbation methods in fluid mechanics", Academic Press, New York (1964).

Jacqueline Sanchez-Hubert

(Laboratoire de Mecanique Theorique

Universite P. et M. Curie - Tour 66

4 place Jussieu, 75230 Paris Cedex 05 - France

Generalised eigensolution expansions in linear acoustics

R.H. PICARD

0. INTRODUCTION

The propagation of acoustic waves is governed by a system of differential equations of the type ($t > 0$, time parameter)

$$\begin{aligned} \operatorname{div} \vec{v}(t) + \frac{\partial}{\partial t} \rho(t) &= f(t) \\ \operatorname{grad} \rho(t) + \frac{\partial}{\partial t} \vec{v}(t) &= \vec{g}(t), \end{aligned} \tag{0.1}$$

$f, \vec{g}$ given, ρ density, $\vec{v}$ velocity vector field. Here all parameters describing the quality and kind of medium are (for sake of simplicity) omitted. The presence of an object is usually modelled as a boundary condition along the surface of the part of space σ occupied by this obstacle. The above system is then considered in $G := \mathbb{R}^3 \setminus \sigma$. The simplest and best known boundary conditions are

$$\vec{n}.\vec{v} = 0 \quad \text{on} \quad \partial G \quad (\vec{n} \text{ denotes the normal vector field on } \partial G) \tag{0.2}$$

and

$$\rho = 0 \quad \text{on} \quad \partial G. \tag{0.3}$$

In order to have a well-posed problem usually an initial condition is imposed:

$$\begin{aligned} \rho(0) &= \rho_o, \\ \vec{v}(0) &= \vec{v}_o, \quad \rho_o, \vec{v}_o \text{ given initial values.} \end{aligned} \tag{0.4}$$

Using the convenient notation of a formal matrix differential operator A

$$A := i \begin{pmatrix} 0 & \operatorname{div} \\ \operatorname{grad} & 0 \end{pmatrix}, \tag{0.5}$$

we get as a Hilbert space formulation of the problem (0.1), (0.3), (0.4) an initial value problem:

$$\frac{\partial}{\partial t} U(t) = i\mathcal{A}\, U(t) + \begin{pmatrix} f(t) \\ \vec{g}(t) \end{pmatrix},$$

$$U(0) = U_o \equiv \begin{pmatrix} \rho_o \\ \vec{v}_o \end{pmatrix}, \tag{0.6}$$

where $\mathcal{A}: D(\mathcal{A}) \subset H \to H$, $U \mapsto AU$, with $H \equiv (L_2(G))^4$ equipped with the natural inner product $(\cdot,\cdot)$ of the direct sum ($L_2(G)$ denotes the usual function space of measurable, square integrable functions) and ('$\ldots^T$' means 'transposed')

$D(\mathcal{A}) := \{(u,v)^T \in L_2(G) \times (L_2(G))^3 \equiv H \mid A(u,v)^T \in H$ and $(\phi, A(u,v)^T) = (A\phi, (u,v)^T)$ for all $\phi \in \{0\} \times (L_2(G))^3 \subset H$ with $A\,\phi \in H\}$

There is a similar transcription for the so-called Newmann case (0.1), (0.2), (0.4). But again to simplify the presentation, we only discuss boundary condition (0.2) (Dirichlet case) and assume furthermore $f \equiv 0$, $\vec{g} \equiv 0$ in G.

For $U_o \in D(\mathcal{A})$ the solution of (0.6) is easy to give, since $\mathcal{A}$ turns out to be self adjoint. We have

$$U(t) = e^{it\mathcal{A}} U_o, \tag{0.7}$$

where $e^{it\mathcal{A}}$ can be interpreted in the sense of spectral theory; thus

$$U(t) = \int_{\mathbb{R}} e^{it\lambda}\, d\Pi(\lambda) U_o, \tag{0.8}$$

where $(\Pi(\lambda))_{\lambda \in \mathbb{R}}$ is the spectral family of $\mathcal{A}$ (see e.g. [2],p.506). Representation (0.8) does not contain more information than the self adjointness of $\mathcal{A}$ and is in so far abstract as it refers to an abstract existence theorem for the spectral family $(\Pi(\lambda))_{\lambda \in \mathbb{R}}$.

The aim of this paper is to show a method that allows us to add 'more meaning' (in a constructive sense) to (0.8).

The basic idea (presented in [6]) is to use the whole space as a reference situation and regard the exterior problem, i.e. the operator $\mathcal{A}$, as a perturbation of the 'full space' problem.

1. THE FULL SPACE PROBLEM

In this case we use the index o to distinguish the reference problem from the perturbed problem, e.g. A_o, $\mathcal{A}_o$, H_o etc.

Using $A_{o,o}(p)$ as a denotation of the normalised symbol of A_o, i.e.

$$A_{o,o}(p) := |p|^{-1} \; F A F^{-1} \; (p),$$

where $(F\phi)(p) := (2\pi)^{-3/2} \int_{\mathbb{R}^3} e^{ix\cdot p}\phi(x)dx$ (Fourier transform applied to every component of $\phi \in H$), we can construct (observing that $A^3_{o,o} = A_{o,o}$)

$$E^{\pm}_o (x,p) := \tfrac{1}{2} (A^2_{o,o}(p) \pm A_{o,o}(p)) \; e^{-ixp} \; (2\pi)^{-3/2} \tag{1.1}$$

as a solution of

$$(A_o \mp |p|) \; E^{\pm}_o (x,p) = 0, \tag{1.2}$$

in the sense of a natural dyadic extension of the meaning of A_o (every row of $E^{\pm}_o (\cdot,p)$ satisfies the equation). Since for every $a_o \in \mathbb{R}^4$ with $E^{\pm}_o (\cdot,p)a_o \not\equiv 0$ we have

$$E^{\pm}_o (\cdot,p)a_o \notin H_o,$$

we say that $E^{\pm}_o (\cdot,p)$ is a <u>generalised</u> eigensolution.
We will employ $E^{\pm}_o$ as a kernel of an integral transform ($\phi \in H_o$):

$$(\xi^{\pm}_o \phi)(p) := \int_{\mathbb{R}^3} \overline{E^{\pm}_o(x,p)} \; \phi(x) \; dx, \; p \in \mathbb{R}^3. \tag{1.3}$$

Using (1.3) we get for the Fourier transform F applied to $\phi \in H_o$, $\phi \perp N(\mathcal{A}_o)$ ('N(...)' means 'null space of ...'; orthogonality $\perp$ in the sense of H_o),

$$F\phi = \xi^+_o\phi + \xi^-_o\phi \; ; \tag{1.4}$$

in other words $\xi^+_o + \xi^-_o$ is projection on the closure of the range $R(\mathcal{A}_o)$ of $\mathcal{A}_o$, i.e. on $\overline{R(\mathcal{A}_o)}$.

Rewriting a representation of the spectral family of $\mathcal{A}_o$ given in [3], we get

$$\Pi_o(\lambda)\phi = \int_{|p|\leq\lambda} E^+_o (\cdot,p) \; (\xi^+_o\phi)(p) \; dp + \int_{-|p|\leq\lambda} E^-_o(\cdot,p)(\xi^-_o\phi)(p) \; dp \tag{1.5}$$

for all $\phi \in \overline{R(\mathcal{A}_o)}$, $\lambda \in \mathbb{R}$.

Furthermore we have Parseval's equality

$$||\Pi_o(\lambda)\phi||^2 = \int_{|p|\leq\lambda} |(\xi^+_o \phi)(p)|^2 dp + \int_{-|p|\leq\lambda} |(\xi^-_o\phi)(p)|^2 \; dp \tag{1.6}$$

for all $\phi \in \overline{R(\mathcal{A}_o)}$.

2. THE PERTURBED PROBLEM

We will now develop a similar representation to (1.5) for the operator $\mathcal{A}$.

This will be done in two steps:

1. Step: Define generalised eigensolutions and corresponding integral transforms.

2. Step: Evaluate Stone's representation formula using Step 1.

Accordingly, we are looking first for generalised eigensolutions of $\mathcal{A}$:

$(A \mp |p|) \underset{\sim}{E}^{\pm}(\cdot,p) = 0$ in G,

$\underset{\sim}{E}^{\pm}(\cdot,p)$ satisfies the boundary condition built-in in $D(\mathcal{A})$,

and $\underset{\sim}{E}^{\pm}(x,p)$ should behave like $E_o^{\pm}(x,p)$ for large $|x|$.

<u>1. Step:</u>

In order to find such a $\underset{\sim}{E}^{\pm}$, we set

$$F^{\pm}(\cdot,p) := A(1-j(\cdot)) \; E_o^{\pm}(\cdot,p), \; p \in \mathbb{R}^3,$$

where $j(x) \in C_{\infty}(\mathbb{R}^3)$ with $j(x) \equiv 1$ outside a compact set and $j(x) \equiv 0$ in a neighbourhood of ∂G. Obviously, $F^{\pm}(\cdot,p)$ has compact support. For $\lambda \in \mathbb{C}$ in the resolvent set $\rho(\mathcal{A}) = \mathbb{C} \setminus \mathbb{R}$ of $\mathcal{A}$, we set

$$\tilde{E}^{\pm}(\cdot,p,\lambda) := (\mathcal{A} - \lambda)^{-1} \; F^{\pm}(\cdot,p), \; p \in \mathbb{R}^3; \text{ then} \tag{2.2}$$

$$E^{\pm}(x,p,\lambda) := j(x) \; E_o^{\pm}(x,p) + \tilde{E}^{\pm}(x,p,\lambda), \; x \in G, \; p \in \mathbb{R}^3, \tag{2.3}$$

satisfies

$$(A \pm |p|) \; E^{\pm}(\cdot,p,\lambda) = (\pm |p| - \lambda) \; j(\cdot) \; E_o^{\pm}(\cdot,p), \; p \in \mathbb{R}^3, \; \lambda \in \rho(\mathcal{A}). \tag{2.4}$$

The limiting process

$$\lambda \to \pm |p| + i0 \; (\text{or } \pm |p| - i0)$$

leads to the desired generalised eigensolution. The existence of a limit in an appropriate sense is assured by the principle of limiting absorption. Boundaries for which the limiting absorption principle is valid have been characterized in [4] and [5], (it is enough to assume the restricted cone property for ∂G; for a definition compare [1], p.11).

We denote our generalised eigensolution by

$$E^{\pm,+}(\cdot,p) := E^{\pm}(\cdot,p,\pm \; |p| + i0) \text{ (and analogously for } \pm |p| - i0), \; p \in \mathbb{R}^3 . \tag{2.5}$$

As corresponding integral transforms we have

$$(\xi^{\pm,\lambda}\phi)(p) := \int_G \overline{E^{\pm}(x,p,\lambda)}\, \phi(x)\, dx, \quad p \in \mathbb{R}^3, \ \lambda \in \mathbb{C}\setminus\mathbb{R}. \qquad (2.6)$$

The limit $\ell m\, \lambda \to 0$ again is controlled by the principle of limiting absorption using detailed consideration along the lines of [6]. In the limit case $\lambda \to \pm |p| + i0$ we will write $\xi^{\pm,+}$ for the corresponding transform.

We have the following useful relation between $\xi^{\pm,\lambda}$ and $\xi_o^{\pm}$ ($f \in H$, supp f compact)

$$\xi^{\pm,\lambda} f(p) = (\pm |p| - \lambda)\, \xi_o^{\pm} (j\, (\mathcal{A}-\lambda)^{-1} f)(p), \ \lambda \in \mathbb{C}\setminus\mathbb{R}, \ p \in \mathbb{R}^3 \qquad (2.7)$$

(compare [6], p.93).
Now we are ready for the

2. Step:

Stone's representation formula expresses the spectral family $(\Pi(\lambda))_{\lambda\in\mathbb{R}}$ of a self adjoint operator $\mathcal{A}$ in terms of its resolvents $R(\mathcal{A}) := (\mathcal{A} - \lambda)^{-1}$, $\lambda \in \rho(\mathcal{A})$.

Assuming that $f = \mathcal{A}w$, $g = \mathcal{A}v$, $w,v \in D(\mathcal{A})$, supp w and supp v compact, we can determine $(f,\Pi(\lambda)g)$, $\lambda \in \mathbb{R}$, (using (2.7) and the limiting absorption principle) as superposition of generalised eigensolutions:
We first have by Stone's representation formula that ($a \in \mathbb{R}$)

$$(f,[\Pi(\lambda) - \Pi(a)]g) = \lim_{\epsilon\to 0+} \int_a^{\lambda} (f,[R(\sigma+i\varepsilon) - R(\sigma-i\varepsilon)]g)\,d\sigma; \qquad (2.8)$$

since the limiting absorption principle implies that $(1-j)\, R(\sigma+i\varepsilon)g$ is uniformly bounded in H and the projection of $jR(\sigma+i\varepsilon)g$ on the null space $N(\mathcal{A}_o)$ of $\mathcal{A}_o$ is uniformly bounded in H_o as $\varepsilon \to 0\pm$, we have from (1.4) for the integrand in (2.8)

$$\begin{aligned}
&(f,[R(\sigma+i\varepsilon) - R(\sigma-i\varepsilon)]g) = 2i\varepsilon\, (R(\sigma+i\varepsilon)f,\, R(\sigma+i\varepsilon)g) \\
&= 2i\varepsilon\, (jR(\sigma+i\varepsilon)f,\, jR(\sigma+i\varepsilon)g) + o(1) \\
&= 2i\varepsilon\, [(\xi_o^{+}(jR(\sigma+i\varepsilon)f),\, \xi_o^{+}(jR(\sigma+i\varepsilon)g)) \\
&+ (\xi_o^{-}(jR(\sigma+i\varepsilon)f),\, \xi_o^{-}(jR(\sigma+i\varepsilon)g))] + o(1).
\end{aligned}$$

Using (2.7) we get for (2.8)

$$(f,[\Pi(\lambda) - \Pi(a)]g) = \lim_{\varepsilon\to 0+} \frac{\varepsilon}{\pi}\int_a^\lambda \{\int_{\mathbb{R}^3} \frac{(\xi^{+,\sigma+i\varepsilon}f)(p)\ \overline{(\xi^{+,\sigma+i\varepsilon}g)(p)}}{(|p| - \sigma)^2 + \varepsilon^2}\, dp$$

$$+ \int_{\mathbb{R}^3} \frac{(\xi^{-,\sigma+i\varepsilon}f)(p)\ \overline{(\xi^{-,\sigma+i\varepsilon}g)(p)}}{(|p| + \sigma)^2 + \varepsilon^2}\, dp\}d\sigma \qquad (2.9)$$

Applying Fubini's theorem and Stieltjes' inversion formula to (2.9) and letting $a \to -\infty$ yields

$$(f,\pi(\lambda)g) = \int_{|p|\leqslant\lambda} (\xi^{+,\pm}f)(p)\ \overline{(\xi^{+,\pm}g)(p)}\ dp + \qquad (2.10)$$

$$+ \int_{|p|\leqslant\lambda} \xi^{-,\pm}f(p)\ \overline{\xi^{-,\pm}g\ (p)}\ dp.$$

A standard completion argument shows that (2.10) is true for all $f,g \in \overline{R(\mathcal{A})}$. Rewriting (2.10) leads to the desired representation (compare (1.5))

$$\Pi(\lambda)g = \int_{|p|\leqslant\lambda} E^{+,\pm}(\cdot,p)(\xi^{+,\pm}g)(p)\ dp +$$

$$+ \int_{-|p|\leqslant\lambda} E^{-,\pm}(\cdot,p)\ (\xi^{-,\pm}g)(p)\ dp \qquad (2.11)$$

for all $g \perp N(\mathcal{A})$, $\lambda \in \mathbb{R}$.

CONCLUDING REMARKS

Though there is not very much new in our arguments (compared to [6]), this simplified presentation gives a model for the treatment of more complicated situations (e.g. variable coefficients) or less well-known initial boundary value problems. Using literally the same sequence of arguments, a corresponding result can be shown e.g. in the case of Maxwell's initial boundary value problem. In fact, in a more general setting the similarities become obvious (compare [3]), thus making the validity of an analogue to (2.11) in this case at least plausible. Finally it should be mentioned that the integral transforms $\xi^{\pm,+}$, $\xi^{\pm,-}$ have the expected property:

$$\xi^{\pm,+}\psi(\mathcal{A}) = \psi(\pm|p|)\ \xi^{\pm,+} \quad \text{(similarly for } \xi^{\pm,-}) \qquad (2.12)$$

for ψ a bounded and Lebesgue measurable function.

Along analogous considerations as in [6] relation (2.12) can be used to derive a representation of the wave operator for $\mathcal{A}$.

LITERATURE

1. S. Agmon: Lectures on Elliptic Boundary Value Problems. Van Nostrand Mathematical Studies 2, Princeton (1965).
2. G. Bachmann, L. Narici: Functional Analysis. Academic Press, New York (1966).
3. R. Picard: Zur Existenz des Wellenoperators bei Anfangsrandwertproblemen vom Maxwell-Typ. Math.Z. 156, 175,185 (1977).
4. C. Weber: Hilbertraummethoden zur Untersuchung der Beugung elektromagnetischer Wellen an Dielektrika. Dissertation, Stuttgart (1977).
5. N. Weck: Maxwell's Boundary Value Problem on Riemannian Manifolds with Nonsmooth Boundaries. J. Math. Anal. and Appl. 46, 410-437 (1974).
6. C.H. Wilcox: Scattering Theory for the d'Alembert Equation in Exterior Domains. Lecture Notes Nr.442, Springer-Verlag, Berlin (1975).

Dr. R. Picard.
Institut fur Angewandte Mathematik,
Universitat Bonn,
Abteilung fur Mathematische Methoden der Physik,
D-5300 Bonn,
Wegelerstrasse 10,
West Germany.

The Rothe method and discontinuities in wave propagation

E. MARTENSEN

1. INTRODUCTION

The Rothe method (or horizontal line method) has for a long time been applied mainly to parabolic evolution problems; as in fact was done by Rothe himself [10]. Recently in the case of hyperbolic problems, the convergence of the Rothe method has been shown for the vibrating elastic string problem by Gerdes and Martensen [1] and for the Maxwell equations and wave equation in several space dimensions by Martensen [5,6]. As these papers are concerned with the approximation of classical solutions by the Rothe method only, the initial data, with respect to the propagation of discontinuities along characteristic lines, had to be assumed with some strong regularity. In another line of development, under considerably more general suppositions, v. Welck [11] and Picard [8,9] have studied the Rothe method independently of the type of the underlying partial differential equation problem. In an analogous general sense, the convergence of the Rothe method is obtained by these authors.

In the following the Rothe method will be investigated with the aim of establishing pointwise convergence for wave propagation in one space dimension containing discontinuities. For this we return to the initial boundary value problem for the vibrating string which we shall prefer to describe as a first order evolution problem. So if the string is of length π, we have the system of differential equations

$$u_t = v_x, \; v_t = u_x \text{ in } [0,\pi]\times[0,\infty) \tag{1}$$

with initial values which we consider to be given by the Fourier series

$$u(x,0) = \sum_{\nu=1}^{\infty} a_\nu \sin \nu x, \quad v(x,0) = \sum_{\nu=0}^{\infty} b_\nu \cos \nu x \quad \text{in } [0,\pi] \tag{2}$$

and with boundary values

$$u(0,t) = 0, \; u(\pi,t) = 0 \text{ in } [0,\infty); \tag{3}$$

here $u(x,t)$, $x\in[0,\pi]$, $t\in(0,\infty)$ denotes the displacement of the string and $v(x,t)$, $x\in[0,\pi]$, $t\in(0,\infty)$, may be interpreted as a velocity potential. The uniquely determined well-known solution of this problem is given by the Fourier series

$$\left.\begin{aligned} u(x,t) &= \sum_{\nu=1}^{\infty} a_\nu \sin \nu x \cos \nu t - \sum_{\nu=1}^{\infty} b_\nu \sin \nu x \sin \nu t \\ v(x,t) &= \sum_{\nu=1}^{\infty} a_\nu \cos \nu x \sin \nu t + \sum_{\nu=0}^{\infty} b_\nu \cos \nu x \cos \nu t \end{aligned}\right\} \text{ in } [0,\pi]\mathrm{x}[0,\infty). \tag{4}$$

If, for any time level $t>0$, there have been carried out n Rothe steps with equidistant step length $\frac{t}{n}>0$, n being an arbitrary natural number, the corresponding Rothe approximations are known to be [1]

$$\left.\begin{aligned} u_n(x,t) &= \sum_{\nu=1}^{\infty} a_\nu \sin \nu x \,\mathrm{Re}\, \frac{1}{(1-\frac{i\nu t}{n})^n} - \sum_{\nu=1}^{\infty} b_\nu \sin \nu x \,\mathrm{Im}\, \frac{1}{(1-\frac{i\nu t}{n})^n} \\ v_n(x,t) &= \sum_{\nu=1}^{\infty} a_\nu \cos \nu x \,\mathrm{Im}\, \frac{1}{(1-\frac{i\nu t}{n})^n} + \sum_{\nu=0}^{\infty} b_\nu \cos \nu x \,\mathrm{Re}\, \frac{1}{(1-\frac{i\nu t}{n})^n} \end{aligned}\right\}$$

$$\text{in } [0,\pi]\mathrm{x}(0,\infty). \tag{5}$$

Thus, for any fixed pair $x\in[0,\pi]$, $t\in(0,\infty)$, the problem of pointwise convergence arises as to whether or not the approximate solution (5) converges as $n\to\infty$ to the exact solution (4).

In all cases where the series (5), with respect to n, are uniformly convergent, the limit as $n\to\infty$ may be interchanged with the sums, so leading immediately to the exact solution (4). In particular this can be done for all classical solutions, i.e. all continuously differentiable solutions of (1) to (3). This follows from the conditions

$$\sum_{\nu=1}^{\infty} |a_\nu| < \infty, \quad \sum_{\nu=0}^{\infty} |b_\nu| < \infty \tag{6}$$

which hold for the Fourier coefficients and lead to convergent majorant series for (5) independent of n [1]. But there are also non-classical solutions, i.e. solutions in the distribution sense of (1) to (3) which by the same argument, may be obtained as the limit of the Rothe solutions (5). Here we only have to think of all continuous states (2) which, after being extended by its Fourier series into the whole real axis, merely have piecewise continuous derivatives; hence (6) remains still valid.

2. THE EXTREMELY PLUCKED ELASTIC STRING

The above conclusion however fails if the uniform convergence of the series (5) either is not valid or not known. This situation is given for initial states (2) containing discontinuities where generally one of the two conditions (6) does not hold. In order to study the behaviour of the Rothe method in such cases, we shall restrict attention to the example of the so called extremely plucked elastic string which is characterised by the initial values

$$u(x,0) = \begin{cases} 0 \ , \ x = 0 \ , \\ 1 \ , \ x \in (0,\pi) \ , \\ 0 \ , \ x = \pi \ , \end{cases} \quad v(x,0) = 0, \ x \in [0,\pi]. \tag{7}$$

In this case the values of the well-known distributional solution of (1) to (3) can be seen in Figure 1.

As the Fourier coefficients for (7) are given by

$$a_\nu = \begin{cases} \dfrac{4}{\nu\pi} \ , \ \nu = 1,3,5, \ldots \ , \\ \\ 0 \ , \ \nu = 2,4,6, \ldots \ , \end{cases} \qquad b_\nu = 0, \ \nu = 0,1,2, \ldots, \tag{8}$$

and if for simplicity, we denote by Σ' a sum taking into consideration odd indices only, then we find from (5) that the corresponding Rothe solutions are

1) In [1] certain twice continuously differentiable solutions (1) to (3) are considered. Consequently on eliminating v(x,t) from (3), the displacement u(x,t) turns out as a classical solution of the wave equation

$u_{tt} = u_{xx}$ in $[0,\pi] \times |0,\infty)$.

$$\left.\begin{aligned} u_n(x,t) &= \frac{2}{\pi} \sum_{\nu=1}^{\infty}{}' \frac{\sin \nu x}{\nu} \left\{ \frac{1}{(1-\frac{i\nu t}{n})^n} + \frac{1}{(1+\frac{i\nu t}{n})^n} \right\} \\ v_n(x,t) &= \frac{2}{\pi i} \sum_{\nu=1}^{\infty}{}' \frac{\cos \nu x}{\nu} \left\{ \frac{1}{(1-\frac{i\nu t}{n})^n} - \frac{1}{(1+\frac{i\nu t}{n})^n} \right\} \end{aligned}\right\} \text{ in } [0,\pi]\times(0,\infty). \quad (9)$$

By numerical computation of these series, Halter and Munz [3] found empirically good convergence towards to exact solution.

In order to investigate the convergence of $u_n(x,t)$ and $v_n(x,t)$, the idea now is to substitute for the series which occur certain improper integrals. This will be done by means of a quadrature formula given essentially as in Goodwin [2] and Martensen [4] and working as follows. Let $f(\xi)$, $\xi\in\mathbb{R}$, be a real-analytic function satisfying the conditions

(i) $f(\xi)$, $\xi\in\mathbb{R}$, for some real number $\sigma>0$, has an analytic extension $f(\xi)$ into the strip

$$B=\{\zeta\in\mathbb{C} \mid \operatorname{Re}\zeta\in\mathbb{R}, \quad \operatorname{Im}\zeta\in[0,\sigma]\}, \tag{10}$$

(ii) $f(\xi+i\sigma)$, for $\xi\in\mathbb{R}$, is absolutely integrable,

(iii) $f(\xi+i\eta)$, with respect to $\eta\in[0,\sigma]$, converges uniformly towards zero, as ξ tends to $-\infty$ and ∞.

Then the given function $f(\xi)$, for $\xi\in\mathbb{R}$, is also integrable and the (improper) integral, for an arbitrarily chosen initial point $\xi_o\in\mathbb{R}$ and any real step length $h>0$, may be calulcated by the rectangular rule

$$\int_{-\infty}^{\infty} f(\xi)d\xi = h \sum_{\nu=-\infty}^{\infty} f(\xi_o+\nu h) + E(\xi_o;h), \tag{11}$$

with derivative-free remainder estimation

$$|E(\xi_o;h)| < (\coth \frac{\pi\sigma}{h} - 1) \int_{-\infty}^{\infty} |f(\xi+i\sigma)|d\xi. \tag{12}$$

As we are interested in pointwise convergence of (9), from now on, $x\in[0,\pi]$ and $t\in(0,\infty)$ will be considered to be fixed. Furthermore, the rectangular rule (11) and (12) will always be applied with

$$\sigma = \frac{\lambda n}{t} > 0, \tag{13}$$

where $\lambda\in(0,1)$ is a suitably chosen real constant, dependent only on x and t.

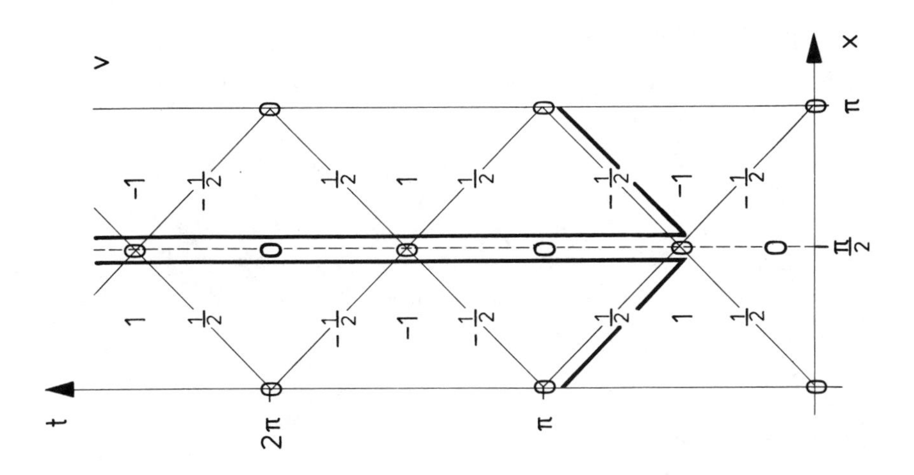

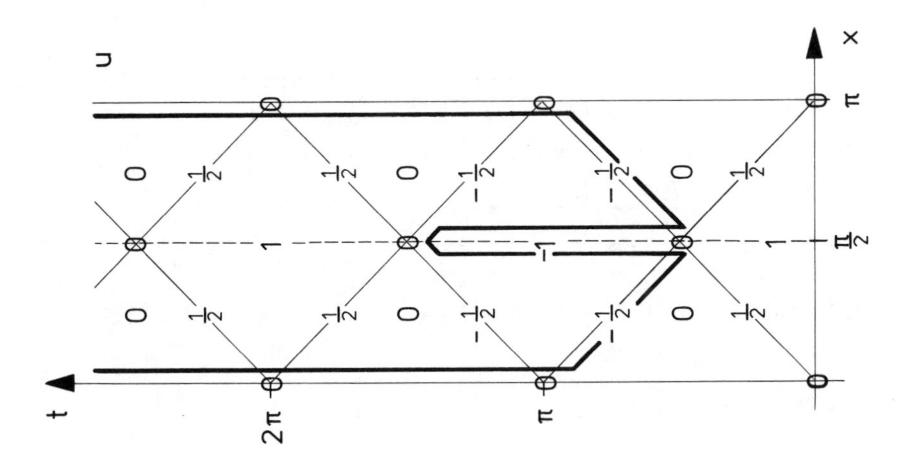

Figure 1. Wave propagation along an extremely plucked elastic string and domains for which the convergence of the Rothe method is proved.

The method will lead to convergence of the Rothe method at all points lying below the thick lines drawn in Figure 1. At all other points the question of convergence remains open.

3. CONVERGENCE OF $u_n(x,t)$

With respect to (9), we define the real-analytic symmetric function sequence

$$f_n(\xi) := \frac{\sin x\xi}{2\pi\ \xi}\left\{\frac{1}{(1-\frac{it\xi}{n})^n} + \frac{1}{(1+\frac{it\xi}{n})^n}\right\},\quad \xi\in\mathbb{R}. \tag{14}$$

Replacing ξ by ζ, we are at once led to the required analytic extension $f_n(\zeta)$, $\zeta\in B$, where the strip $B\subset\mathbb{C}$ is given by (10) and (13). From (14), together with the elementary inequality

$$\left|\frac{\sin z}{z}\right| \leqslant \frac{\sinh \operatorname{Im} z}{\operatorname{Im} z},\quad z\in\mathbb{C}, \tag{15}$$

we obtain for all $\zeta=\xi+i\eta\in B$

$$|f_n(\zeta)|\leqslant\frac{x}{2\pi}\ \frac{\sinh x\eta}{x\eta}\left|\frac{1}{1-\frac{it\zeta}{n}} + \frac{1}{1+\frac{it\zeta}{n}}\right|\left|\sum_{\nu=0}^{n-1}\frac{(-1)^\nu}{(1-\frac{it\zeta}{n})^{n-1-\nu}\ (1+\frac{it\zeta}{n})^\nu}\right|$$

$$\leqslant \frac{x}{\pi}\ \frac{\sinh x\sigma}{x\sigma}\left|\frac{1}{1+\frac{t^2}{n^2}\zeta^2}\right|\sum_{\nu=0}^{n-1}\frac{1}{\left|1-\frac{it\zeta}{n}\right|^{n-1-\nu}\left|1+\frac{it\zeta}{n}\right|^\nu}$$

$$\leqslant \frac{\sinh x\sigma}{\pi\sigma}\ \frac{1}{1+\frac{t^2}{n^2}(\xi^2-\sigma^2)}\ \frac{n}{(1-\frac{t}{n}\sigma)^{n-1}}$$

and hence because of (13) the estimate

$$|f_n(\zeta)|\leqslant\frac{t}{\lambda\pi}\ \frac{\sinh\frac{\lambda nx}{t}}{(1-\lambda)^{n-1}}\ \frac{1}{(1-\lambda^2+\frac{t^2}{n^2}\xi^2},\quad \zeta\in B \tag{16}$$

It is now easily seen, that the function (14) satisfies conditions (i), (ii), and (iii), so we are in the position to apply the rectangular rule. Furthermore, as a consequence of (16) we get the estimate

$$\int_{-\infty}^{\infty}|f_n(\xi+i\sigma)|\,d\xi<\frac{n}{\lambda}\sqrt{\frac{1-\lambda}{1+\lambda}}\ \frac{\sinh\frac{\lambda nx}{t}}{(1-\lambda)^n}\ . \tag{17}$$

Taking now $\xi_o=1$ and $h=2$ and, as a consequence of (iii) and the symmetry of (14), observing that

$$2\sum_{\nu=-\infty}^{\infty} f_n(1+2\nu) = 2\sum_{\nu=-\infty}^{\infty}{}' f_n(\nu) = 4\sum_{\nu=1}^{\infty}{}' f_n(\nu), \tag{18}$$

we get, together with (11) and (14),

$$J_n := \int_{-\infty}^{\infty} f_n(\xi)d\xi = \frac{2}{\pi}\sum_{\nu=1}^{\infty}{}' \frac{\sin \nu x}{\nu}\left\{\frac{1}{(1-\frac{i\nu t}{n})^n} + \frac{1}{(1+\frac{i\nu t}{n})^n}\right\} + E_n(1;2). \tag{19}$$

We shall now evaluate the inproper integral on the left in (19) which by symmetry, may be expressed as the Cauchy mean value integral

$$J_n = \frac{1}{2\pi i}\int_{-\infty}^{\infty} \frac{e^{ix\xi}-1}{\xi}\left\{\frac{1}{(1-\frac{it\xi}{n})^n} + \frac{1}{(1+\frac{it\xi}{n})^n}\right\} d\xi . \tag{20}$$

With regard to the residue method (Figure 2), we consider the analytic function

$$g_n(\zeta) := \frac{e^{ix\zeta}-1}{\zeta}\left\{\frac{1}{(1-\frac{it\zeta}{n})^n} + \frac{1}{(1+\frac{it\zeta}{n})^n}\right\} \tag{21}$$

in the closed upper half plan except the point

$$\zeta_o := \frac{in}{t} . \tag{22}$$

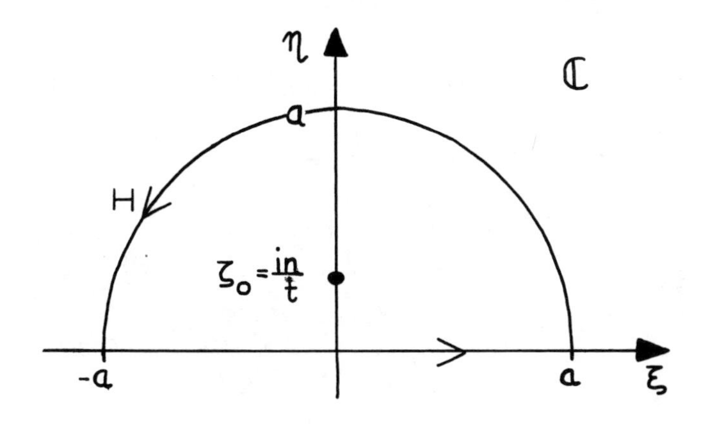

Figure 2. Application of the residue method.

On the semi circle (Figure 2)

$$H : \zeta = ae^{i\phi}, \quad 0 \leqslant \phi \leqslant \pi \;, \tag{23}$$

with radius $a > \frac{n}{t}$ we get the estimates

$$|e^{ix\zeta}| = |e^{-ax\sin\phi + iax\cos\phi}| = e^{-ax\sin\phi} \leqslant 1, \tag{24}$$

$$\left|1 \mp \frac{it\zeta}{n}\right| = \left|1 \pm \frac{at}{n}\sin\phi \mp \frac{iat}{n}\cos\phi\right| = \sqrt{1 \pm \frac{2at}{n}\sin\phi + \frac{a^2t^2}{n^2}} \geqslant \frac{at}{n} - 1 > 0. \tag{25}$$

From these, together with (21), it follows that

$$\left|\int_H g_n(\zeta)d\zeta\right| \leqslant \int_H |g_n(\zeta)|ds \leqslant \int_H \frac{4}{a}\frac{1}{(\frac{at}{n}-1)^n}ds = \frac{4\pi}{(\frac{at}{n}-1)^n} \;. \tag{26}$$

Thus, as $a \to \infty$, the integral over H tends to zero. As can be seen from Figure 2, we may compute the integral (20) as the residue of the function (21) with respect to the singularity (22):

$$J_n = \frac{1}{(n-1)!}\left.\frac{d^{n-1}}{d\zeta^{n-1}}\frac{e^{ix\zeta}-1}{\zeta}\frac{1}{(\frac{it}{n})^n}\right|_{\zeta=\zeta_o}$$

$$= \frac{1}{(n-1)!}\left(-\frac{in}{t}\right)^n\left\{\sum_{\nu=0}^{n-1}\binom{n-1}{\nu}\left(\frac{1}{\zeta}\right)^{(n-1-\nu)}\left(e^{ix\zeta}\right)^{(\nu)} - \left(\frac{1}{\zeta}\right)^{(n-1)}\right\}\Bigg|_{\zeta=\zeta_o}$$

$$= (-\zeta_o)^n\left\{\sum_{\nu=0}^{n-1}\frac{(-1)^{n-\nu-1}}{\nu!}\frac{1}{\zeta^{n-\nu}}(ix)^\nu e^{ix\zeta} - \frac{(-1)^{n-1}}{\zeta^n}\right\}\Bigg|_{\zeta=\zeta_o}$$

$$= \sum_{\nu=0}^{n-1}\frac{(-1)^{-\nu-1}}{\nu!}(ix\zeta_o)^\nu e^{ix\zeta_o} + 1 = 1 - e^{-\frac{nx}{t}}\sum_{\nu=0}^{n-1}\frac{1}{\nu!}\left(\frac{nx}{t}\right)^\nu. \tag{27}$$

From this, due to methods of Perron [7], we obtain the limit

$$\lim_{n\to\infty} J_n = \begin{cases} 1 \;, & t<x \;, \\ \frac{1}{2} \;, & t=x \;, \\ 0 \;, & t>x \;. \end{cases} \tag{28}$$

For the remainder in (19), from (12), (13), and (17) it follows that

$$|E_n(1;2)| \leqslant \frac{n}{\lambda} \left(\coth \frac{\lambda n\pi}{2t} - 1\right) \frac{\sinh \frac{\lambda nx}{t}}{(1-\lambda)^n} \tag{29}$$

$$= \frac{n}{\lambda} \frac{2e^{-\frac{\lambda n\pi}{t}}}{1-e^{-\frac{\lambda n\pi}{t}}} \frac{e^{\frac{\lambda nx}{t}} - e^{-\frac{\lambda nx}{t}}}{2} e^{-n\ln(1-\lambda)} \leqslant \frac{n}{\lambda} \frac{e^{-\lambda n\left(\frac{\pi-x}{t} - 1 - \frac{\lambda}{2} - \frac{\lambda^2}{3} - \ldots\right)}}{1-e^{-\frac{\lambda n\pi}{t}}} .$$

If we now restrict attention to such x and t for which

$$0<t<\pi-x, \tag{30}$$

then we can choose $\lambda\in(0,1)$ dependent on x and t only and satisfying the condition

$$\frac{\pi-x}{t} - 1 - \frac{\lambda}{2} - \frac{\lambda^2}{3} - \ldots > 0 . \tag{31}$$

After this, we find from (29) the limit

$$\lim_{n\to\infty} E_n(1;2) = 0. \tag{32}$$

Due to (9), (19), (28), and (32), we see that in the case of (30), we have obtained the required convergence of $u_n(x,t)$. By symmetry of $u_n(x,t)$ with respect to $x = \frac{\pi}{2}$, this result immediately also holds for

$$0<t<x \tag{33}$$

(c.f. (Figure 1)).

The case $x = \frac{\pi}{2}$ and $t\in(0,\infty)$ will be discussed separately. The rectangular formula (11), again applied to the function (14) but now taken for $\xi_o=0$ and $h=1$, yields

$$J_n = f_n(0) + 2 \sum_{\nu=1}^{\infty} f_n(\nu) + E_n(0;1)$$

$$= \frac{1}{2} + \frac{1}{\pi} \sum_{\nu=1}^{\infty}{}' \frac{\sin \frac{\nu\pi}{2}}{\nu} \left\{ \frac{1}{\left(1-\frac{i\nu t}{n}\right)^n} + \frac{1}{\left(1+\frac{i\nu t}{n}\right)^n} \right\} + E_n(0;1). \tag{34}$$

From (12), (13), and (17), we get the remainder estimate

$$|E_n(0;1)| \leqslant \frac{n}{\lambda}\left(\coth\frac{\lambda n\pi}{t}-1\right)\frac{\sinh\frac{\lambda n\pi}{2t}}{(1-\lambda)^n}$$

$$= \frac{n}{\lambda}\,\frac{2e^{-\frac{2\lambda n\pi}{t}}}{1-e^{-\frac{2\lambda n\pi}{t}}}\,\frac{e^{\frac{\lambda n\pi}{2t}}-e^{-\frac{\lambda n\pi}{2t}}}{2}\,e^{-n\ln(1-\lambda)}$$

$$\leqslant \frac{n}{\lambda}\,\frac{e^{-\lambda n\left(\frac{3\pi}{2t}-1-\frac{\lambda}{2}-\frac{\lambda^2}{3}-\dots\right)}}{1-e^{-\frac{2\lambda n\pi}{t}}}\,. \tag{35}$$

By an analogous treatment, after restricting t to

$$0<t<\frac{3\pi}{2}\;, \tag{36}$$

we get for a suitable choice of $\lambda\epsilon(0,1)$

$$\lim_{n\to\infty} E_n(0;1) = 0. \tag{37}$$

This together with (9), (28), and (34) yields the required result; namely $u_n(\frac{\pi}{2},t)$, for $t\epsilon(0,\frac{\pi}{2})$, $t=\frac{\pi}{2}$, $t\epsilon(\frac{\pi}{2},\frac{3\pi}{2})$, converges to 1,0,-1, respectively (Figure 1).

Finally we mention the trivial fact that $u_n(0,t)$ and $u_n(\pi,t)$, for all $t\epsilon(0,\infty)$, both converge to 0 (Figure 1).

4. CONVERGENCE OF $v_n(x,t)$

Here we start by considering the real-analytic symmetric function

$$f_n(\xi) := \frac{\cos x\xi}{2\pi i\xi}\left\{\frac{1}{(1-\frac{it\xi}{n})^n}-\frac{1}{(1+\frac{it\xi}{n})^n}\right\}\;,\;\xi\epsilon\mathbb{R}\;, \tag{38}$$

with corresponding analytic extension $f_n(\zeta)$, $\zeta=\xi+i\eta\epsilon B$. Because of

$$|\cos z| \leqslant \cosh \operatorname{Im} z\;,\; z\epsilon\mathbb{C}\;, \tag{39}$$

and (13), we find

$$|f_n(\zeta)| \leqslant \frac{\cosh x\eta}{2|\zeta|}\left|\frac{1}{1-\frac{it\zeta}{n}}-\frac{1}{1+\frac{it\zeta}{n}}\right|\left|\sum_{\nu=0}^{n-1}\frac{1}{(1-\frac{it\zeta}{n})^{n-1-\nu}(1+\frac{it\zeta}{n})^{\nu}}\right|$$

$$\leqslant \frac{t\cosh x\sigma}{n\pi}\,\frac{1}{1+\frac{t^2}{n}(\xi^2-\sigma^2)}\,\frac{n}{(1-\frac{t}{n}\sigma)^{n-1}}$$

$$=\frac{t}{\pi}\,\frac{\cosh\frac{\lambda nx}{t}}{(1-\lambda)^{n-1}}\,\frac{1}{1-\lambda^2+\frac{t^2}{n^2}\xi^2}\,,\quad \zeta\epsilon B, \tag{40}$$

and further

$$\int_{-\infty}^{\infty}|f_n(\xi+i\sigma)|\,d\xi \leqslant n\sqrt{\frac{1-\lambda}{1+\lambda}}\,\frac{\cosh\frac{\lambda nx}{t}}{(1-\lambda)^n}\,. \tag{41}$$

Hence, the application of the rectangular rule (11) to (28) is allowed again and, taking there $\xi_o=1$ and $h=2$, together with (18) we have

$$J_n:=\int_{-\infty}^{\infty}f_n(\xi)d\xi=\frac{2}{\pi i}\sum_{\nu=1}^{\infty}{}'\frac{\cos\nu x}{\nu}\left\{\frac{1}{(1-\frac{i\nu t}{n})^n}-\frac{1}{(1+\frac{i\nu t}{n})^n}\right\}+E_n(1;2). \tag{42}$$

For the residue method, with regard to (38) we may write J_n as the Cauchy mean value integral

$$J_n=\frac{1}{2\pi i}\int_{-\infty}^{\infty}\frac{e^{ix\xi}}{\xi}\left\{\frac{1}{(1-\frac{it\xi}{n})^n}-\frac{1}{(1+\frac{it\xi}{n})^n}\right\}d\xi. \tag{43}$$

As above the corresponding integral over the semi circle H vanishes if the radius tends to infinity. Thus (43) can be calculated as the residue of the integrand with respect to the singularity ζ_o (Figure 2). Referring to the computation already done in (27), we obtain

$$J_n=-\frac{1}{(n-1)!}\frac{d^{n-1}}{d^{\,n-1}}\frac{e^{ix\zeta}}{\zeta}\frac{1}{(\frac{it}{n})^n}\Bigg|_{\zeta=\zeta_o}=e^{-\frac{nx}{t}}\sum_{\nu=0}^{n-1}\frac{1}{\nu!}\left(\frac{nx}{t}\right)^{\nu} \tag{44}$$

and from this the limit

$$\lim_{n\to\infty} J_n = \begin{cases} 0 & , \quad t\ x \quad , \\ \frac{1}{2} & , \quad t=x \quad , \\ 1 & , \quad t>x \quad . \end{cases} \tag{45}$$

For the remainder in (42), because of (12), (13), and (41), we obtain

$$|E_n(1;2)| \leqslant n \left(\coth \frac{\lambda n\pi}{2t} - 1\right) \frac{\cosh \frac{\lambda nx}{t}}{(1-\lambda)^n} . \tag{46}$$

Thus analogously to the above, for all x and t satisfying (30), it follows that

$$\lim_{n\to\infty} E_n(1;2)=0 . \tag{47}$$

Now from (9), (42), (45), and (47), it follows that $v_n(x,t)$ if restricted by (30), converges towards the exact solution. Because of the anti-symmetry of $v_n(x,t)$ with respect to $x=\frac{\pi}{2}$, the same result is obtained under the condition (33). The convergence of $v_n(\frac{\pi}{2},t)$ to 0, for all $t\in(0,\infty)$, is trivial (Figure 1).

REFERENCES

[1] W. Gerdes und E. Martensen. Das Rothe-Verfahren für die räumlich eindimensionale Wellengleichung. Z.Angew.Math.Mech. 58 (1978), T367-T368.

[2] E.T. Goodwin. The evaluation of integrals of the form $\int_{-\infty}^{\infty} f(x)e^{-x^2}dx$. Proc.Cambr.Phil.Soc. 45 (1949), 241-245.

[3] E. Halter und C.-D. Munz. Hyperbolische Evolutionsprobleme mit unstetigen Daten und ihre Lösung mit der horizontalen Linienmethode. To appear in Z.Angew.Math.Mech.

[4] E. Martensen. Zur numerischen Auswertung uneigenlicher Integrale Z.Angew.Math.Mech. 48 (1968), T83-T85.

[5] E. Martensen. The convergence of the horizontal line method for Maxwell's equations. Math.Methods Appl.Sci. 1 (1979), 101-113.

[6] E. Martensen. The Rothe method for the wave equation in several space dimensions. Proc.Roy.Soc.Edin. 84A (1979), 1-18.

[7] O. Perron. Über die näherungsweise Berechnung von Funktionen grosser Zahlen. Sitzungsberichte der math.-phys. Klasse der bayerischen Akademie der Wissenschaften 1917, 191-219.

[8] R. Picard. On convergence and error estimates for the horizontal line method applied to Maxwell's initial boundary value problem in anisentropic, inhomogeneous media. Proc.Roy. Soc.Edin. 86A (1980), 53-59.

[9] R. Picard. Some remarks on the horizontal line method. Math. Methods Appl.Sci. 2 (1980), 471-479.

[10] E. Rothe. Zweidimensionale parabolische Randwertaufgaben als Grenzfall ein-dimensionaler Randwertaufgaben. Math.Ann. 102 (1930), 650-670.

[11] U.v. Welck. Ein stabiles Schichtenverfahren für allgemeine lineare Evolutionsgleichungen. Numer.Math. 27 (1977), 171-178.

Professor Dr. E. Martensen,
Mathematisches Institut II,
Universität Karlsruhe (TH),
D-75 Karlsruhe,
West Germany.

Spectral properties of differential equations

D. PEARSON

I believe that this contribution to the proceedings will be unique in that it is motivated by problems in quantum mechanics rather in classical mechanics. Nevertheless, in hearing from other contributors of their concern with eigenvectors of self-adjoint and non-self-adjoint operators, with the location of the essential spectrum, with studying the singularities of the S-Matrix, and so on, it is easy to convince oneself that the two disciplines are not so dissimilar after all. In this spirit I shall present a treatment of some aspects of spectral theory of the Schrödinger operator.

We shall restrict our attention to a very simple class of operators, which illustrate many of the ideas involved. Let $V(r)$ be a real-valued potential function, defined for $0 \leqslant r < \infty$. We shall suppose that V is locally square integrable, regular near $r=0$; but no restriction will be made on the behaviour of V for large values of r. The Hamiltonian operator $\hat{H} = -\frac{d^2}{dr^2} + V(r)$ may be defined with domain $D(\hat{H}) = C_0^\infty(0,\infty)$. Let H be a self-adjoint extension of $\hat{H}$, acting in $L^2(0,\infty)$. We shall suppose that we have the <u>limit point case</u> at $r=\infty$. In other words, no boundary condition at infinity is required, and H may be determined by a single boundary condition (typically $\phi(0)=0$ at $r=0$, and we shall assume that this condition is imposed).

We are interested in the spectral properties of the self-adjoint operator H, and in particular in how these are related to the behaviour of solutions $u(r,z)$ of the time-independent Schrödinger equation

$$- \frac{d^2u}{dr^2} + Vu = zu \qquad (z\in\mathbb{C}). \tag{1}$$

Let u_1, u_2 respectively, be the solutions of (1) satisfying the initial conditions

$$\left.\begin{array}{ll} u_1(0,z) = 0 & u_2(0,z) = 1 \\ (\frac{du_1}{dr})(0,z) = 1 & \frac{du_2}{dr}(0,z) = 0 \end{array}\right\}. \tag{2}$$

For any value of z such that Imz>0, there is a solution u of (1), unique up to a multiplicative constant, such that $u(.,z)\in L^2(0,\infty)$. Since this solution cannot be a multiple of u_1 (otherwise the self-adjoint operator H would have a complex eigenvalue), we can define, for Imz>0, a function m(z) by the property

$$u_2(.,z) + m(z)u_1(.,z) \in L^2(0,\infty). \tag{3}$$

The function m(z) is <u>analytic</u> in z (for Imz>0) and has positive imaginary part. In fact, one has an explicit formula for the imaginary part of m(z), in terms of the L^2 norm of u(.,z), viz.

$$\frac{\text{Im.}m(z)}{\text{Im}z} = \int_0^\infty |u_2(r,z) + m(z)u_1(r,z)|^2 dr. \tag{4}$$

Now, any function m(z) analytic in the upper half plane, and having positive imaginary part, may be represented in the form

$$m(z) = \int_{-\infty}^{\infty} \frac{d\rho(\lambda)}{(\lambda - z)}, \tag{5}$$

where $\rho(\lambda)$ is a non-decreasing function, continuous from the right, and unique up to an additive constant, for given m(z). (Strictly speaking this is not quite true, since (i) we could add to the right hand side a positive multiple of z, and (ii) subtractions may be necessary in order to make the integral converge. However, in the present context m(z) behaves sufficiently well for large $|z|$ that we need make only a single subtraction, and (5) becomes $m(z) - m(z_o) = \int_{-\infty}^{\infty} (\frac{1}{\lambda - z} - \frac{1}{\lambda - z_o}) d\rho(\lambda)$. For simplicity, we shall continue to use (5), which will hold, for example, under the assumption $\rho(\infty) - \rho(-\infty)$ is finite).

The function $\rho(\lambda)$ generates a Stieltjes measure μ on the Borel subsets of the real line, such that, for intervals $(a,b]$, $\mu(a,b] = \rho(b) - \rho(a)$.

Thus, to any Schrödinger operator $H = -\frac{d^2}{dr^2} + V$ in $L^2(0,\infty)$ one may associate a unique measure μ. The importance of this measure μ lies in the fact that the operator H is unitarily equivalent to the operator of multiplication by λ (i.e. $h(\lambda) \rightsquigarrow \lambda h(\lambda)$) in the space $L^2(\mathbb{R}; d\mu)$. Hence a knowledge of the measure μ allows us to make a complete spectral analysis of H.

Given an arbitrary measure μ on the Borel subsets of $\mathbb{R}$, there is a decomposition $\mu = \mu_{a.c.} + \mu_s$ of μ into its absolutely continuous and singular parts. A measure γ is said to be absolutely continuous if $\gamma(A)=0$ for every set A having Borel measure zero. Such a measure has a density function f which is locally integrable, and unique in the sense that any two density functions differ on a set of (Borel) measure zero. Thus if γ is absolutely continuous we have, for arbitrary A,

$$\gamma(A) = \int_A f(\lambda)d\lambda.$$

On the other hand, γ is said to be singular if γ is concentrated on a set of Borel measure zero.

The singular part μ_s of μ may be decomposed still further into the discrete and singular continuous parts, thus $\mu_s = \mu_d + \mu_{s.c.}$. Here μ_d is concentrated on a discrete, countable set of points (the discrete points of the measure), each of which has strictly positive μ-measure. $\mu_{s.c.}$, on the other hand, has no discrete points, and is either identically zero or is concentrated on an uncountable set of points having Borel measure zero.

Further decompositions of the measure are being found to be of importance in quantum mechanics, but we shall not discuss them further here.

How can the properties of the measure μ generated by $\rho(\lambda)$ be derived from a knowledge of the function $m(z)$ in (5).

Setting $z = x+iy$, we have

$$\operatorname{Im} m(z) = \int_{-\infty}^{\infty} \frac{y\, d\rho(\lambda)}{(\lambda-x)^2+y^2}. \tag{6}$$

For fixed λ, we have $\lim\limits_{y\to 0+} \frac{y}{(\lambda-x)^2+y^2} = 0$ for $\lambda \neq x$, and

$\int_{-\infty}^{\infty} \frac{y}{(\lambda-x)^2+y^2}\, dx = \pi$. Hence, for small positive y the function

$\frac{y}{(\lambda-x)^2+y^2}$ looks like $\pi\delta(\lambda-x)$, where δ is a Dirac delta function.

If the measure μ were purely absolutely continuous, with density $f(\lambda)$, we could write $d\rho(\lambda) = f(\lambda)d\lambda$ in the integral on the right hand side of (6), which leads us to expect that, in that case, $\mathrm{Im}\, m(z)$ approaches $\pi d\rho/d\lambda$ in the limit as $y \to 0+$. This intuition may be generalised to give a correct formula, if we remember that, if ρ is differentiable, we have

$$\frac{d\rho(\lambda)}{d\lambda} = \lim_{\delta\to 0+} \frac{\rho(\lambda+\delta)-\rho(\lambda-\delta)}{2\delta} = \lim_{\delta\to 0+} \frac{\mu(\lambda-\delta,\lambda+\delta]}{2\delta} .$$

Quite generally, that is for an arbitrary measure μ, we have

$$\lim_{y\to 0+} \mathrm{Im}.m(x+iy) \equiv \lim_{y\to 0+} \int_{-\infty}^{\infty} \frac{y\, d\rho(\lambda)}{(\lambda-x)^2+y^2} = \lim_{\delta\to 0+} \frac{\pi\mu(x-\delta,x+\delta]}{2\delta} . \qquad (7)$$

Equation (7) holds in the sense that the two limits are equal whenever either limit exists. Actually, we know that the limits _do_ exist, for almost all x, this being a consequence of the result that an analytic function in the upper half plane, having positive imaginary part, has a boundary value almost everywhere along the real z axis.

There is another sense in which $m(z)$ has a boundary value as z approaches the real axis. Namely, $\lim_{y\to 0+} m(x+iy)$ exists _in the sense of distributions_, and the limit is just $\pi\mu$. (Here $\pi\mu$ stands for the distribution $\phi \leadsto \pi\int \phi d\mu$, for $\phi \in C_0^\infty(\mathbb{R})$). But equation (7) enables us to consider _pointwise_ boundary values of $m(z)$, and to relate these to properties of the measure μ. Equation (7) may also be used to study the spectral properties of an arbitrary self-adjoint operator T, in a Hilbert space $\mathcal{H}$ since if T has the representation $T = \int_{-\infty}^{\infty} \lambda dE_\lambda$, and if f is an arbitrary vector in $\mathcal{H}$, the relationship (5) between a function $m(z)$ analytic in the upper half plane, with positive imaginary part, and a measure $\mu = d\rho(\lambda)$ holds if we set

$$m(z) = \langle f,(T-z)^{-1}f\rangle \quad \text{and} \quad \rho(\lambda) = \langle f,E_\lambda f\rangle .$$

In that case it is clear that, in considering boundary values of

$m(z)$, we are actually studying the behaviour near the real axis of the resolvent operator $(T-z)^{-1}$. In this connection, it should be noted that the decomposition of the measure μ corresponds to the decomposition $\mathcal{H} = \mathcal{H}_d \oplus \mathcal{H}_{a.c.} \oplus \mathcal{H}_{s.c.}$ of the Hilbert space, where a vector f belongs, for example, to the subspace $\mathcal{H}_{a.c.}$ if and only if the measure generated by the function $\langle f, E_\lambda f \rangle$ is absolutely continuous. The space $\mathcal{H}_{a.c.}$ is particularly important in quantum scattering theory, since one may show, for a particle moving in a short range (but possibly singular) potential, that states which, for large positive and negative times, are located far from the scattering centre, always belong to the subspace of absolute continuity of the total Hamiltonian H. The subspace $\mathcal{H}_d$, on the other hand, is the subspace spanned by all of the eigenvectors, and the subspace $\mathcal{H}_{s.c.}(H)$ contains the "localised" continuum states.

Returning now to equation (7), the following measure theoretical results allow us to interpret the right hand side

1. Let the set $\mathcal{S}$ consist of all $x \in \mathbb{R}$ such that $\frac{\mu(x-\delta, x+\delta]}{2\delta}$ is bounded as δ approaches 0. (The bound will depend, in general, on the particular value of x). Then the restriction of μ to $\mathcal{S}$, which we denote by $\mu|_{\mathcal{S}}$, is absolutely continuous, and indeed $\mu|_{\mathcal{S}} = \mu_{a.c.}$. Moreover, the density $f(\lambda)$ of $\mu_{a.c.}$ is given, for almost all λ, by

$$f(\lambda) = \lim_{\delta \to 0+} \frac{\mu(\lambda-\delta, \lambda+\delta]}{2\delta}.$$

2. The restriction of μ to the complement of $\mathcal{S}$ is the singular part of the measure; $\mu|_{\mathcal{S}^c} = \mu_s$. Moreover, a measure μ is singular if and only if $\lim_{\delta \to 0+} \frac{\mu(x-\delta, x+\delta]}{2\delta} = 0$ for almost all $x \in \mathbb{R}$.

Results 1 and 2 are the key to the study of the spectral properties of $H = -\frac{d^2}{dr^2} + V(r)$ in terms of the boundary values of $m(z)$. In connection with 1, one may prove a result analogous to (7), namely that $\frac{\mu(x-\delta, x+\delta]}{2\delta}$ is bounded as $\delta \to 0+$ if and only if $\mathrm{Im}.m(x+iy)$ is bounded as $y \to 0+$. So the correspondence between the behaviour of μ and the behaviour of Im.m is very close indeed.

There is, however, one key to spectral analysis that needs to be introduced. It is all very well to try to analyse the behaviour of m(z) as z approaches the real axis. But the definition of m(z), through equation (3), involves in practise a detailed knowledge of solutions u(r,z) of equation (1) for complex values of z, and this knowledge is often extremely difficult to obtain. In order to reduce the problem, in certain respects, to that of deriving information concerning behaviour of solutions $u(r,\lambda)$ of the differential equation

$$-\frac{d^2u}{dr^2} + Vu = \lambda u, \tag{8}$$

for <u>real</u> values of λ, we introduce the notion of <u>subordinacy</u>.

<u>Definition</u>: A non-zero solution $u(r,\lambda)$ of (8) is said to be <u>subordinate</u>, for a given value of λ, if, for any linearly independent solution $v(r,\lambda)$ of (8), we have

$$\lim_{N\to\infty} \frac{\| u(.,\lambda)\|_N}{\| v(.,\lambda)\|_N} = 0, \tag{9}$$

where $\|.\|_N$ denotes a norm in the Hilbert space $L^2(0,N)$.

<u>Examples</u>: (i) Let $u(r,\lambda)$ be any $L^2(0,\infty)$ solution of (8). Then $u(r,\lambda)$ is subordinate, since (in the first point casr at $r=\infty$) only one such solution (apart from constant multiples of it) can exist.

(ii) Let $u_o(r,\lambda)$ be any non-oscillating solution of (8). For example, suppose $u_o(r,\lambda)>0$ for r sufficiently large. Then if $v(r,\lambda)$ is any solution of (8), we have $\frac{d}{dr}\{\frac{v(r,\lambda)}{u_o(r,\lambda)}\} = \frac{u_o v' - u_o' v}{u_o^2} = \frac{\text{const}}{u_o^2}$, which has fixed sign.

Hence $\lim_{r\to\infty} \frac{v(r,\lambda)}{u_o(r,\lambda)} = \alpha$ exists as an extended real number (that is, $\alpha\in\mathbb{R}$, or $\alpha = \pm\infty$).

If $|\alpha|=\infty$, then $u_o(r,\lambda)$ is subordinate. If $\alpha\in\mathbb{R}$, then $v(r,\lambda)-\alpha u_o(r,\lambda)$ is subordinate.

Thus non-oscillating solutions always imply subordinacy.

The use of subordinacy is well illustrated by the following <u>Theorem</u> (D. Gilbert).

Suppose that, for some real λ, $m(\lambda)\equiv\lim_{y\to 0+} m(\lambda+iy)$ exists and the limit is real. Then $u_2(r,\lambda)+m(\lambda)u_1(r,\lambda)$ is a subordinate solution of (8), where u_1, u_2 satisfy (8) with the initial conditions (z) (λ replacing z).

Conversely, suppose a (real) subordinate solution exists, say $u_2(r,\lambda)+M(\lambda)u_1(r,\lambda)$. Then either Im.$m(\lambda+iy)$ is unbounded as $y\to 0+$, or $\lim_{y\to 0+} m(\lambda+iy)=M(\lambda)$, (Take $M(\lambda)=\infty$ if $u_1(r,\lambda)$ is subordinate).

As applications of these ideas, we shall give just two fairly straightforward consequences of the above Theorem.

Application 1: Suppose that, for λ in the neighbourhood of some point λ_o, there exists a subordinate solution of equation (8) which belongs to $L^2(0,\infty)$. Then, either λ_o is not in the spectrum of H, or λ_o is an isolated eigenvalue of H, or λ_o is a limit point of H.

Idea of proof: Let the subordinate solution be

$$u(r,\lambda) = u_2(r,\lambda) + M(\lambda)u_1(r,\lambda).$$

(either $M(\lambda)\in\mathbb{R}$, or adopt the convention $M(\lambda)=\infty$ if u_1 is subordinate.) If $M(\lambda)=\infty$ then λ is an eigenvalue of H, and to prove that result it is only necessary to consider the case where $M(\lambda)\in\mathbb{R}$ for λ sufficiently close to λ_o. In that case, for given, fixed $\phi\in C_0^\infty(0,\infty)$, one has $(H-\lambda)\psi=\phi$, where

$$\psi(r;\lambda)=u_1(r,\lambda)\int_r^\infty (u_2(r',\lambda)+M(\lambda)u_1(r',\lambda))\phi(r')dr'$$

$$+(u_2(r,\lambda)+M(\lambda)u_1(r,\lambda))\int_0^r u_1(r',\lambda)\phi(r')dr'.$$

In particular, $(H-\lambda)^{-1}\phi$ exists for λ close to λ_o, so that $\|(H-\lambda)^{-1}\phi\|^2 = \int_{-\infty}^{\infty} (\lambda-\mu)^{-2}d\langle\phi, E_\mu\phi\rangle <\infty$ for λ close to λ_o, say for λ belong to some neighbourhood $\mathcal{N}$ of λ_o. This is sufficient to imply that $\int_{\mathcal{N}} d\langle\phi, E_\mu\phi\rangle=0$, and since this holds for a dense set of $\phi\in L^2(0,\infty)$ it follows in this case that $\int_{\mathcal{N}} dE_\mu=0$, and consequently that λ_o does not lie in the spectrum of H.

Application 2: Suppose that, for almost all λ in the neighbourhood of some point λ_o, equation (8) has a subordinate solution which does not belong to $L^2(0,\infty)$. Then the neighbourhood of λ_o is contained in the singular spectrum of H.

Idea of a proof: Since, for a dense set of λ near λ_o, there is no solution of (8) in $L^2(0,\infty)$, standard results imply that these values of λ belong to the spectrum of H. Since the spectrum is closed, the entire neighbourhood is contained in the spectrum. For almost all λ, $m(\lambda+iy)$ approaches a limit as $y\to0+$, in which case certainly $\text{Im}.m(\lambda+iy)$ is bounded in this limit. Hence, by the Theorem, $\lim_{y\to0+} m(\lambda+iy)$ exists and is real for almost all λ near λ_o. Comparing result 2 above with equation (7), it follows that the spectrum of H in this region is singular.

Although in these applications we have dealt in more detail with the singular spectrum of H, comparable results will hold for the absolutely continuous spectrum. Of course the problem remains, for given $V(r)$, to obtain sufficiently detailed estimates of the asymptotics of solutions to equations (8) in order to apply the general theory. All methods of spectral analysis in these problems have to concern themselves with a study of these asymptotics. The following examples illustrate some of the interesting spectral behaviour which can occur, in cases which it has been possible to analyse by one method or another. They should be compared with the "standard" example of spectral properties, that of a potential $V(r)$ of the form $V=V_1+V_2$, where $V_1\in L'(0,\infty)$ and V_2 is of bounded variation, with $V_2(r)\to0$ as $r\to\infty$. In that case $H=-\frac{d^2}{dr^2}+V(r)$ has absolutely continuous spectrum for $\lambda\geqslant0$ and discrete spectrum (isolated eigenvalues, possibly with an accumulation point at $\lambda=0$).

Example 1: $V(r) = \cos(e^r)$.

Spectral properties the same as the "standard" example. (The oscillations of V are too rapid to have much effect on asymptotic behaviour).

Example 2: $V(r) = e^r\cos(e^r)$.

The same as the "standard" example, except that the absolutely continuous spectrum starts at $\lambda=-\frac{1}{2}$. (The value $\frac{1}{2}$ is the mean squared value of the cosine unction!).

Example 3: $V(r) = \sum_1^\infty a_n \cos(r/2^n)$, with $\sum_1^\infty |a_n| < \infty$.

For a sense set of $\{a_n\}$ (i.e. dense in the sequence space ℓ') the spectrum is absolutely continuous, but a Cantor set.

One can also have singular continuous spectrum. The simplest example of this is not a differential operator but a matrix, as follows:

Example 4: Let K be the infinite matrix with components $[K]_{ij} = \delta_{i,j+1} + \delta_{i+1,j}$. (K is the discrete analogue of $-\frac{d^2}{dx^2}$).

Let V be the infinite diagonal matrix with components $[V]_{ij} = \beta\cos(2\pi\alpha i + \theta)\,\delta_{ij}$, where α is a Liouville number and $\beta>2$. Then K+V, regarded as a matrix operator in $\ell^{(2)}$, has singular continuous spectrum for almost all values of θ.

REFERENCES

A good reference for the introductory material is [1]. The Theorem is due to D. Gilbert (to be published) who also contributed to other developments. For an example of singular continuous spectrum of Schrödinger operators see [2]. Examples 3 and 4 at the end of this article are due to Avron and Simon ([3], [4]). Example 2 is due to the present author.

1. E. Coddington and N. Levison, Theory of ordinary differential equations; New York, McGraw Hill 1955.
2. D.B. Pearson, Comm. Math. Phys. 60, 13 (1978).
3. J. Avron and B. Simon, Almost periodic Schrödinger operators I, Calif. Inst. of Tech. preprint.
4. J. Avron and B. Simon, Singular continuous spectrum for a class of almost periodic Jacobi matrices, Princeton University preprint.

Dr. D.B. Pearson,
Department of Applied Mathematics,
University of Hull,
Hull.

On exterior initial boundary value problems in thermoelasticity

R. LEIS

In [1], [2] initial boundary value problems in thermoelasticity were solved and in the special case of $G=R^3$ the asymptotic behaviour of the solutions was discussed. It is the aim of the following note to treat a special exterior initial boundary value problem and to give the asymptotic behaviour in that case.

Let $G \subset R^3$ be an exterior domain and let the underlying medium be isotropic and homogeneous. Let γ, μ, ν be positive constants. The the equations of thermoelasticity read

$$\ddot{U} + \mu \operatorname{rot} \operatorname{rot} U - \nu \operatorname{grad} \operatorname{div} U + \gamma \operatorname{grad} \theta = 0$$

$$\dot{\theta} - \Delta\theta + \gamma \operatorname{div} \dot{U} = 0.$$

The initial conditions are $U(0)=U^o$, $\dot{U}(0)=U^1$, $\theta(0)=\theta^o$ and for boundary conditions we choose (classically written)

$$n \times U|\partial G = 0, \ \operatorname{div} U|\partial G = 0, \ \theta|\partial G = 0.$$

Thus we are looking for a $U \in C_2(R_o^+, \mathcal{L}_2) \cap C_1(R_o^+, \overset{\circ\circ}{D})$ and a $\theta \in C_1(R_o^+, \mathcal{L}_2)$ with $U \in \overset{\circ}{R} \cap \overset{\circ\circ}{D}$, $\theta \in \overset{\circ}{H}_1$, rct rot $U \in \mathcal{L}_2$ and $\Delta\theta \in \mathcal{L}_2$. We have used

$$\overset{\circ}{R} = \overline{\{\overset{\circ}{C}: \|\cdot\|_r = \sqrt{\|\cdot\|^2 + \|\operatorname{rot} \cdot\|^2}\}}, \quad \overset{\circ\circ}{D} = \{U \in \mathcal{L}_2 | \operatorname{div} U \in \overset{\circ}{H}_1\}.$$

Because of the special structure of our equations, let us project

$$\mathcal{L}_2 = \nabla \overset{\circ}{H}_1 \oplus D_o$$

$$U = U^p + U^s.$$

We then get

$$\operatorname{rot} \operatorname{rot} U = \operatorname{rot} \operatorname{rot} U^s \in D_o$$

$$\operatorname{grad} \operatorname{div} U = \operatorname{grad} \operatorname{div} U^p \in \nabla \overset{\circ}{H}_1$$

$$\operatorname{grad} \theta \in \nabla \overset{\circ}{H}_1$$

and our equations split up into

$$\ddot{U}^s + \mu \operatorname{rot} \operatorname{rot} U^s = 0, \ U^s \in \overset{\circ}{R} \cap D_o \tag{1}$$

and

$$\ddot{U}^p - \nu \text{ grad div } U^p + \gamma \text{ grad } \theta = 0, \quad U^p \in \overset{\circ\circ}{D} \cap \overline{\nabla \overset{\circ}{H}_1}$$
$$\dot{\theta} - \Delta\theta + \gamma \text{ div } \dot{U}^p = 0, \qquad \theta \in \overset{\circ}{H}_1. \tag{2}$$

Equations (1) are the usual Maxwell's equations. One knows that they are uniquely solvable, the solutions have constant energy and for large t, they converge to an asymptotic wave field $U^s_\infty(t,x)$.

Thus let us deal with problem (2). We assume $\nu=1$, div $U^o \in \mathcal{L}_2$ and set

$$V = (\text{div } U^p, \dot{U}^p, \theta)' \quad .$$

Problem (2) then is equivalent to

$$\dot{V} + AV = 0, \; V(0) = V^o = (\text{div } U^{op}, U^{1p}, \theta^o)' \in \mathcal{L}$$

with

$$\mathcal{L} = \mathcal{L}_2 \times \overline{\nabla \overset{\circ}{H}_1} \times \mathcal{L}_2, \; D = \{U \in \mathcal{L}_2 | \text{div } U \in \mathcal{L}_2\}$$
$$D(A) = \{V \in \mathcal{L} | V \in \overset{\circ}{H}_1 \times D \times \overset{\circ}{H}_1 \wedge \Delta V_3 \in \mathcal{L}_2\}$$

and

$$A : D(A) \subset \mathcal{L} \to \mathcal{L}$$
$$V \to \begin{pmatrix} 0 & -\text{div} & 0 \\ -\text{grad} & 0 & \gamma \text{ grad} \\ 0 & \gamma \text{ div} & -\Delta \end{pmatrix} V.$$

A is a closed non selfadjoint operator with $D(A^*) = D(A)$ and $(AV,V) = \|\nabla V_3\|^2 + 2i \text{ Im } \{(V_1, \text{div } V_2) + \gamma(\nabla V_3, V_2)\}$.

Thus for $t \geqslant 0$ a continuous semigroup $H(t)$ exists (compare [2]) with $H(0) = I$, $\|H(t)\| \leqslant 1$, $AH(t) \supset H(t)A$ and

$$V(t) = H(t)V^o, \; V^o \in D(A)$$

is the uniquely determined solution of problem (2). We want to show $\lim_{t\to\infty} \|V(t)\| = 0$.

First we remark that V is a solution with "finite energy"

$$E(t) = (V,V) \leqslant \|V^o\|^2 = E(0). \tag{3}$$

We easily get for all $V^o \in D(A)$

$$\|\dot{V}\| = \|AV\| \leqslant \|AV^o\| \tag{4}$$

and together with

$$\|\nabla V_3\|^2 = -(V_3, \Delta V_3) = (V_3, AV|_3 + \gamma \, AV|_1) \leqslant c\|V^o\| \cdot \|AV^o\|$$

this means

$$V^o \in D(A) \Rightarrow \|\nabla V_1\| + \|\text{div } V_2\| + \|\nabla V_3\| + \|\Delta V_3\|$$
$$\leq c\ \{\|V^o\| + \|AV^o\|\}. \tag{5}$$

Now let us start with

$$E'(t) = 2\ \text{Re}(V,\dot{V}) = -\ 2\ \text{Re}(V,AV) = -\ 2\|\nabla V_3\|^2$$

which gives

$$0 \leq E(t) = E(0) - 2\int_0^t \|\nabla V_3(\tau)\|^2\ d\tau \leq E(0). \tag{6}$$

We want to draw some conclusions out of this. To do so we use

Lemma: $f \in C_1(R_o^+,R_o^+) \cap \mathcal{L}_1(R_o^+,R_o^+) \wedge |f'| \leq \text{const}$

$$\lim_{t\to\infty} f(t) = 0$$

and define

$$\overline{V}(t) := H(t)\overline{V}^o \text{ for } \overline{V}^o := A^{-1}\ V^o,\ V^o \in D(A^{-1})$$

$$\underline{V}(t) := H(t)\underline{V}^o \text{ for } \underline{V}^o := AV^o \quad , V^o \in D(A)$$

which yields

$$A\overline{V} = V,\ \underline{V} = AV,\ \dot{\overline{V}} + A\overline{V} = 0 \text{ and } \dot{\underline{V}} + A\underline{V} = 0 \text{ for } V^o \in D(A^2).$$

Now take $f = \|\nabla V_3\|^2$ and let $V^o \in D(A)$. Then we get

$$f(t+h)-f(t) = -\ (\Delta V_3(t+h),V_3(t+h)-V_3(t))-(V_3(t+h)-V_3(t),\Delta V_3(t))$$

yielding

$$|f'(t)| = -\ 2\ \text{Re}(\Delta V_3,\dot{V}_3) \leq c\ \|AV^o\|\{\|V^o\| + \|AV^o\|\}.$$

Thus equation (6) and the lemma give for $t \to \infty$

$$V^o \in D(A) \Rightarrow \|\nabla V_3\| \to 0. \tag{7}$$

To illustrate our technique we estimate

$$\|V_3\|^2 = (V_3,A\overline{V}|_3) = (V_3,\gamma\,\text{div } \overline{V}_2-\Delta\overline{V}_3) = (\nabla V_3,-\gamma\overline{V}_2+\nabla\overline{V}_3)$$
$$\leq c\ \|\nabla V_3\|\{\|\overline{V}_2\| + \|\nabla\overline{V}_3\|\}$$

which means

$$V^o \in D(A) \cap D(A^{-1}) \Rightarrow \|V_3\| \to 0 \tag{8}$$

In a similar way we get

$$V^o \in D(A^2) \Rightarrow \|\dot{V}_3\| + \|\gamma\,\text{div } V_2-\Delta V_3\| + \|\nabla\dot{V}_3\| \to 0. \tag{9}$$

Finally, let $V^o \in D(A^2)$. Then $AV=HAV^o \in D(A)$ which means div $V^2 \in \overset{o}{H}_1$

yielding

$$(\text{div } V_2, \Delta V_3) = -(\nabla \text{ div } V_2, \nabla V_3) = (\nabla \dot{V}_1, \nabla V_3) = (\nabla \underline{V}_1, \nabla V_3)$$
$$\leq c \; \|\nabla V_3\| \to 0.$$

Thus we get

$$V^o \in D(A^2) \Rightarrow \|\text{div}\, V_2\| + \|\Delta V_3\| \to 0. \tag{10}$$

Now we come to our conclusion. Let $V^o \in D(A) \cap D(A^{-1})$. Then we have

$$\|V_1\|^2 = (V_1, A\overline{V}|_1) = (V_1, -\text{ div } \overline{V}_2) \leq \|V^o\| \, \|\text{ div } \overline{V}_2\| \to 0$$

and similarly for $V^o \in D(A^2) \cap D(A^{-1})$

$$\|V_2\|^2 = (V_2, A\overline{V}|_2) = (V_2, -\nabla\overline{V}_1 + \gamma\nabla\overline{V}_3) = (\text{div } V_2, \overline{V}_1 - \gamma\overline{V}_3) \to 0.$$

Thus we have proved

$$V^o \in D(A^2) \cap D(A^{-1}) \Rightarrow \|V(t)\| \to 0. \tag{11}$$

We remark that equation (11) may be proved for $V^o \in D(A) \cap D(A^{-2})$ also and that

$$V^o \in D(A) \cap D(A^{-2}) \Rightarrow \|U(t)\| \to 0$$

holds. Similarly, if we use the projection

$$\mathcal{L}_2 = \overline{\text{rot } \overset{o}{R}} \oplus R_o$$

we may treat the boundary conditions

$$n\, U|\partial G = 0, \; n \times \text{rot } U|\partial G = 0, \; \theta|\partial G = 0.$$

REFERENCES

[1] Leis, R.: Außenraumaufgaben in der linearen Elastizitätstheorie. Math. Meth. in the Appl. Sci. 2 (1980) 379-396.

[2] Leis, R.: Über das asymptotische Verhalten thermoelastischer Wellen im R^3. Math. Meth. in the Appl. Sci. 3 (1981) 312-317.

[3] Wickel, W.: On initial boundary value problems for the equations of thermoelasticity. Preprint SFB 72. Inst. f. Angewandte Mathematik, Univ. Bonn 1981.

Professor Dr. Rolf Leis,
Institut für Angewandte Mathematik der Universität Bonn,
Wegelerstr. 10, D-5300 Bonn 1

Convergence of geometrical optics approximations if tangential rays of higher order are present

H.-D. ALBER

Let $B = \mathbb{R}^2$ denote a bounded domain with smooth boundary ∂B, and let Ω be the complement of $\bar{B}$. Points in $\mathbb{R}^2$ will be denoted by $x = (x_1, x_2)$. $\mu = e^{-ikx2}$ is a plane wave solution of the equation $\Delta\mu + k^2\mu = 0$, $k^2 > 0$, in $\mathbb{R}^2$. I want to study the reflection of this plane wave at B, and to study the high frequency asymptotics of the reflected wave. Thus, let u be the solution of

$$\Delta u + k^2 u = 0 \text{ in } \Omega,\ k^2 > 0$$

$$u|_{\partial B} = -e^{-ikx_2}$$

$$\frac{\partial}{\partial r} u = iku + O(r^{-\frac{1}{2}}),\ r = |x| \to \infty.$$

I want to show that u can be approximated for large $|k|$ by the Ansatz of geometrical optics

$$u_{GO}(x,k) = e^{ik\phi(x)} \sum_{m=0}^{\infty} k^{-m} z_m(x)$$

with suitable functions ϕ and z_n. Let me review some facts about this Ansatz. The functions ϕ and z_m are determined in the following way. The series defining the Ansatz is inserted in the Helmholtz equation, and then the coefficients of the powers of k are equated to zero. This yields a recursive system of first order partial differential equations for ϕ and z_m

$$|\nabla\phi|^2 = 1$$

$$2\nabla\phi.\nabla z_m + (\Delta\phi)z_m = -\Delta z_{m-1},\ m = 0,1,2, \ldots ,$$

where z_{-1} is defined to be zero. The first equation is called eikonal equation. The second equations are called transport equations.

Also the boundary condition should be satisfied. This yields the following conditions for ϕ and z_m at ∂B.

$$\left.\begin{aligned} &\phi(x) = -x_2, \\ &z_m(x) = \begin{cases} -1, & m = 0 \\ 0, & m > 0 \end{cases} \end{aligned}\right\} x \in \partial B .$$

It is easy to solve this system with the usual theory of partial differential equations of the first order. The characteristics of the eikonal equation can be interpreted as reflected rays:

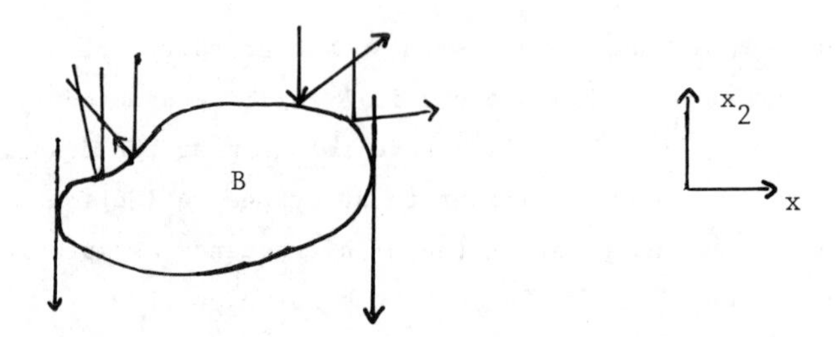

The surfaces ϕ = const are the reflected wave fronts. If ϕ is determined from the eikonal equation, the transport equations are differential equations for the unknown functions z_m, which can be solved recursively. It is well known, however, that the series $e^{ik\phi}\Sigma k^{-m} z_m$ obtained by this procedure in general is not convergent. The solution u thus cannot be represented by this series. Instead, one tries to show that u is represented asymptotically by this series

$$u(x,k) = e^{ik\phi(x)} \sum_{m=0}^{n} k^{-m} z_m(x) + O(k^{-n}), \quad |k| \to \infty.$$

Another problem with this Ansatz is obvious from this picture. If the characteristics of the eikonal equation intersect, in other words, if a caustic is present in the reflected field, the solutions ϕ and z_m cannot be continued across this caustic. A still more difficult problem in dealing with this Ansatz are the tangential rays. On tangential rays the solution ϕ is not two times differentiable, and thus the transport equations have coefficients which are singular on tangential rays.

In the past few years it has been observed that convergence of the Ansatz of geometrical optics to the exact solution is a local problem, and simple proofs for convergence in neighbourhoods of rays being not tangential to the boundary have been found, using the wave equation. On tangential rays, however, things are more difficult. In this talk I want to deal with this problem.

To overcome the difficulties with tangential rays Ludwig suggested a new Ansatz for the approximation of u containing Airy functions. Subsequently Morawetz and Ludwig, and later Taylor and Melrose proved that this Ansatz indeed can be used to approximate the solution on tangential rays. The problem with this new Ansatz is, however, that it only works if the curvature of the boundary ∂B does not vanish. If the curvature vanishes, then difficulties similar to the difficulties with the usual Ansatz appear, and I think that a completely new Ansatz is necessary to handle this case. As far as I know, up to now nobody was able to suggest what this new Ansatz should be. But this is a quite unsatisfactory situation, since one wants to have something which always converges to the exact solution, regardless of whether tangential rays are present or not, and regardless of whether the curvature of B vanishes or not. So I tried to weaken the concept of convergence. Instead of looking for an Ansatz which converges to the solution pointwise-uniformly, I tried to show that the usual Ansatz of geometrical optics converges to the solution in the L_2-sense. I shall now state the results and show how these results can be used to determine the most singular terms of solutions to the wave equation. At the end of the talk I want to explain the idea of proof.

I think that the method also works for a ray tangent to B at a concave part of the boundary.

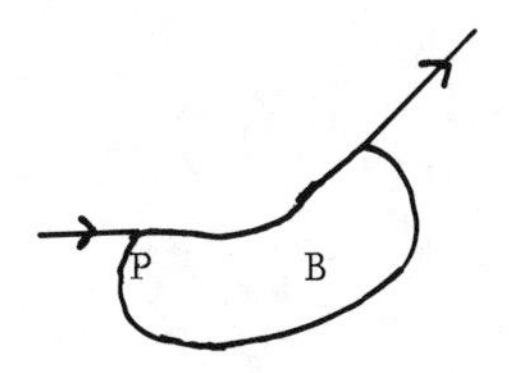

However, since I am still working on this problem, I want to restrict myself in this talk to rays tangent to ∂B at convex parts of the boundary ∂B. So let me assume that B is strictly convex, for technical convenience. This is a real restriction for ∂B only in a neighbourhood of the point of tangency, since the convergence of the geometrical optics approximation is a local problem, as mentioned before, and it is known how to deal with non convex boundaries if no tangential rays are present.

Let $P \in \partial B$. I choose the coordinate system with origin in P as follows. Let the x-axis point in the direction of the normal vector from Ω to B, and let the x_2- axis point in the direction of the vector tangent to B at P. $\mu(x) = e^{-ikx_2}$ is a plane wave with propation direction tangent to ∂B at P, and at a second point $\tilde{P} \in \partial B$. The solution u will be approximated by

$$u_{GO}(x,k) = e^{ik\phi(x)} z_o(x)$$

where ϕ and z_o are solutions of

$$|\nabla\phi|^2 = 1$$

$$\phi(x) = - x_2, \quad x \in \partial B$$

$$\phi(x) \to \infty, \quad |x| \to \infty$$

and

$$2\nabla\phi\cdot\nabla z_o + (\Delta\phi) z_o = 0$$

$$z_o(x) = -1, \; x \in \partial B.$$

It follows from these equations that we have in the shadow S behind the obstacle

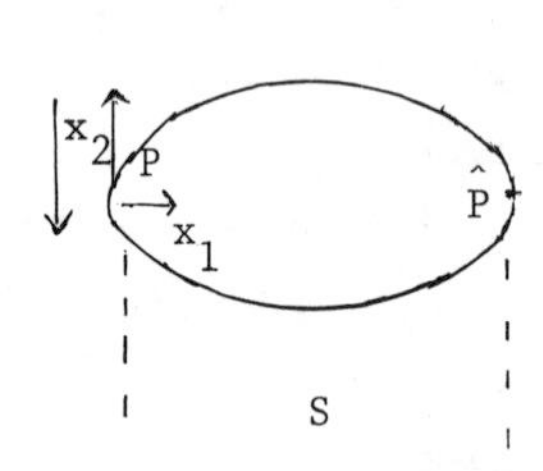

$$u_{GO}(x,k) = -e^{-ikx_2} = -\mu(x), \; x \in S ,$$

hence $u_{GO}(x,k) + \mu(x) = 0, \; x \in S.$

The first result is

Theorem 1: There is a function $|k| \mapsto C(|k|) > 0$, with $C(|k|) > 0$ for $|k| \to \infty$, such that

$$||u(\cdot,k)-u_{GO}(\cdot,k)||_{\Omega_R} \leqslant (1+\varepsilon^{-1/2})R^{3/2+\varepsilon/2}C(|k|)||u||_{\Omega_R} ,$$

for every $\varepsilon > 0$ and $R > 0$, where

$$||u||^2_{\Omega_R} = \int_{\substack{x\in\Omega \\ |x|<R}} |u|^2 dx.$$

For this result I only need the assumption that B is strictly convex. For example, the curvature of ∂B may vanish at P to infinite order. However, the function $C(|k|)$, which gives the convergence speed, depends on the shape of the boundary in a neighbourhood of the points P and $\tilde{P}$. This function decreases to zero rather slowly. If the curvature of ∂B vanishes in both points P and $\tilde{P}$ of finite order, then it is easy to give an estimate for $C(|k|)$. I have the following result.

Theorem 2: Let S denote the arc length of ∂B, and assume that the curvature $\chi(S)$ of ∂B vanishes at P of order $n < \infty$, and at $\tilde{P}$ of order $\tilde{n} < \infty$. Then.

$$||u(\cdot,k)-u_{GO}(\cdot,k)||_{\Omega_R} \leqslant C(1+\varepsilon^{-1/2})R^{3/2+\varepsilon/2}(k^{-\frac{1}{3n+7}}+k^{\frac{1}{3\tilde{n}+7}})||u||_{\Omega_R}.$$

These results can be used to determine the most singular terms of solutions to the wave eqution. Let $S(x,t)$ be a distribution solution of the problem

$$(\partial_t^2-\Delta)S(\lambda,t) = 0 \quad \text{in } \Omega$$

$$S_{|\partial B\times\mathbb{R}} = 0$$

$$S(x,t) = \delta(t+x_2), \text{ t sufficiently negative.}$$

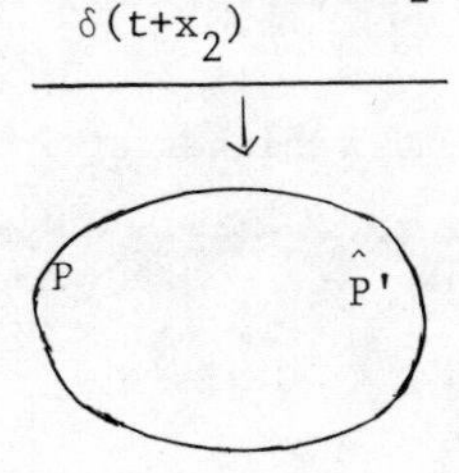

S represents a plane wave front propagating in the direction of the negative x_2-axis and scattered at ∂B. S contains singularities on the rays tangential to ∂B at P and at $\tilde{P}$. The problem is to determine the singularities of S after reflection at ∂B. First I shall give a more precise definition of the solution S of this problem.

Let $\mathcal{S}_o = \mathcal{S}_o(\Omega\times\mathbb{R})$ denote the space of all the functions $\psi \in C^\infty(\bar{\Omega}\times\mathbb{R})$ with $\Delta^m\psi_{|\partial B\times\mathbb{R}} = 0$, $m = 0,1,2, \ldots$, and

$$p_{\alpha m}(\psi) = \sup_{z\in\Omega\times\mathbb{R}} (1+|z|)^m|D^\alpha\psi(z)| < \infty$$

for all multi-indices $\alpha \in \mathbb{N}_o^3$ and $m = 0,1,2, \ldots$. Let $\mathcal{S}_o'$

denote the dual of $\mathcal{S}_o$. For $u \in \mathcal{S}'_o$. I define $\partial_t u$ and Δu by

$$(\partial_t u,\psi) = -(u,\partial_t\psi)$$

and

$$(\Delta u,\psi) = (u,\Delta\psi).$$

I look for a solution $S \in \mathcal{S}'_o$ of the problem

$$(\partial_t^2-\Delta)S = 0$$

and

$S = \delta(t+x_2)$, t sufficiently negative. The boundary condition need not be formulated explicitly since it is implied in the definition of Δ on $\mathcal{S}'_o$. Note that boundary terms are neglected in this definition.

This problem has a unique solution S. It is possible to define the Fourier transform $\hat{S}(x,k)$ of $S(x,t)$ with respect to the t-variable as usual for tempered distributions, and it can be shown that

$$\hat{S}(x,k) = e^{-ikx_2} + u(x,k)$$

for $k \neq 0$, where u is the solution at $\Delta u + k^2u = 0$ in Ω, $u_{|\partial B} = -e^{-ikx_2}$, $\frac{\partial}{\partial r} u = iku + O(r^{-1/2})$, hence u is the function considered in the first part of my talk. These results allow us to show that the inverse Fourier transform $\check{u}_{GO}$ with respect to k of the approximation u_{GO} of u gives the most singular term of S. Thus I have the following result.

<u>Theorem 3</u>: Assume that the curvature $x(s)$ of ∂B vanishes at P of finite order n. Then we have

$$\begin{aligned} S(x,t) &= \delta(t+x_2) + \check{u}_{GO}(x,t) + W(x,t) \\ &= \delta(t+x_2) + \delta(t-\phi(x))z_o(x) + W(x,t) , \end{aligned}$$

with

$$(t\mapsto W(\cdot,t))\in L_2(\mathbb{R},\overset{o}{H}_{-1/2+\eta}(\Omega_R)), \text{ for all } \eta < \frac{1}{3n+7} .$$

ϕ and z_o are the solutions of the eikonal and transport equations. Note that $x \mapsto \delta(t+x_2) + \delta(t-\phi(x))z_o(x)$ is contained in $H_{-1/2}$, but not in $H_{-1/2+\varepsilon}$ for $\varepsilon > 0$, and hence is more singular than $W(x,t)$. $\delta(t+x_2) + \delta(t-\phi(x))z_o(x)$ is a distribution with support on the reflected wave front $\phi = t$.

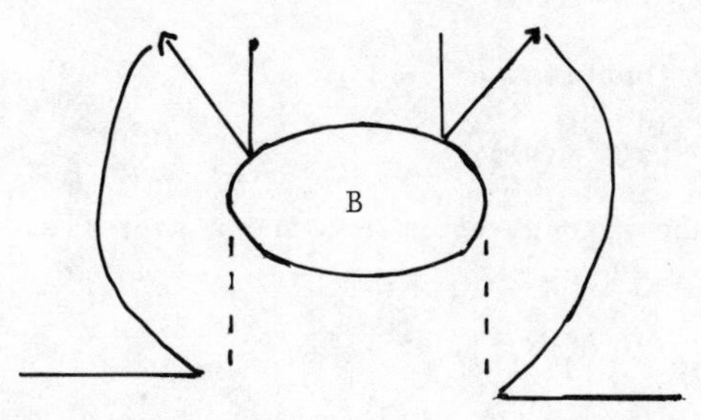

Let me finally describe the idea of proof. Let v be an outgoing solution of

$$\Delta v + k^2 v = \phi, \quad k^2 > 0, \text{ in } \Omega,$$

where $\partial\Omega$ is such that there exist $\beta > 0$ and a suitable coordinate system with

$$-(x \cdot n(x)) \geqslant \beta$$

for all $x \in \partial\Omega$. $u(x)$ is the normal to $\partial\Omega$ at x pointing from Ω to the complement $B = \mathbb{R}^2 \setminus \Omega$. This condition is satisfied if B is convex, for example. In this case I have the following uniform estimate for v. There is a constant $C > 0$ such that for all $R, \varepsilon > 0$

$$||\nabla v||^2_{\Omega_R} + k^2||v||^2_{\Omega_R} \leq CR^{1+\varepsilon}\left[(1+\varepsilon^{-1})|||x|f||^2_{\Omega} + \right.$$

$$\left. + ||\partial_\tau v||^2_{\partial B} + (1+k^2)||v||^2_{\partial B}\right] \tag{1}$$

$\partial_q v$ denotes derivation in the direction of the tangent to ∂B. I derived this inequality using another inequality of Morawetz and Ludwig. For the proof of theorem 1 the difference $u = u_{GO}$, has to be estimated. This difference satisfies the equation

$$(\Delta+k^2)(u-u_{GO}) = -(\Delta+k^2)u_{GO} = -e^{ik\phi}\Delta z_o$$

$$u - u_{GO}|_{\partial B} = 0.$$

The inequality (1) would yield

$$||u-u_{GO}||_{\partial_R} \leqslant C|k|^{-1}|| \, |x|\Delta z_o||_{\Omega} = O(|k|^{-1}), \; |k| \to \infty,$$

since Δz_o is independent of k, by definition. However, this simple idea does not work, since Δz_o is not locally square integrable, due to the singular behaviour of Δz_o on tangential rays. So the idea has to be modified, and some technical details must be inserted.

I introduce the function

$$u_\chi(x,k) = e^{ik\phi(x)} z(x,\gamma(k)),$$

where ϕ satisfies the eikonal equation, and z satisfies

$$2\nabla\phi\cdot\nabla z + (\nabla\phi)z = 0 \text{ in } \Omega$$

$$z(x,\gamma) = -\chi(\gamma,x_1)\chi(\gamma(P_1-x_1)), \quad x \in \partial B,$$

with $\chi \in C^\infty(\mathbb{R})$,

$$\chi(t) = \begin{cases} 0 , & t < 1/2 \\ 1 , & t > 1 \end{cases}$$

and $\gamma > 0$. I shall choose the constant γ depending on k, such that $\gamma(k) \to +\infty$ for $|k| \to \infty$. This construction implies that z vanishes in a neighbourhood of the tangential rays, and is equal to z_o outside another neighbourhood. These neighbourhoods shrink to the tangential rays if $\gamma \to +\infty$. Hence Δz is nonsingular on the tangential rays.

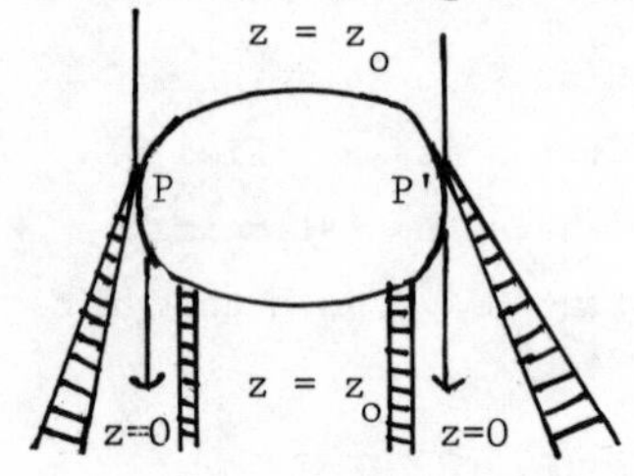

Now the difference

$$||u(\cdot,k)-u_{GO}(\cdot,k)||_{\Omega_R} \leq ||u(\cdot,k)-u_\chi(\cdot,k)||_{\Omega_R} + ||u_\chi(\cdot,k)-u_{GO}(\cdot,k)||_{\Omega_R}$$

can be estimated. It is easy to see that $||u_\chi(\cdot,k) - u_{GO}(\cdot,k)||_{\Omega_R} \to 0$ for $\gamma(k) \to +\infty$. The difference $v = u - u_\chi$ satisfies

$$(\Delta+k^2)v = -(\Delta+k^2)u_\chi = -e^{ik\phi}\Delta z(x,\gamma(k))$$

$$v|_{\partial B} = u - u_\chi|_{\partial B} \quad ,$$

and hence can be estimated with the inequality (1):

$$||v||_{\partial_R}$$

$$\leq C[|k|^{-1}|| \, |x|\Delta z(\cdot,\gamma(k))||_\Omega + |k|^{-1}||\partial_\tau(u-u_x)||_{\partial B} + ||u-u_x||_{\partial B}].$$

It is easy to see that $|k|^{-1}||\partial_\tau(u-u_\chi)||_{\partial B} + ||u-u_\chi||_{\partial B} \to 0$ for $|k| \to \infty$, $\gamma(k) \to +\infty$, since u_χ differs from u at the boundary only in a small neighbourhood at the points P and $\hat{P}$, which shrinks to the points P and $\hat{P}$ for $\gamma(k) \to +\infty$. So it remains to consider the term $|k|^{-1} || |x| \Delta z(\cdot,\gamma(k)) ||_{\Omega_R}$. If γ is fixed, then $|| |x| \Delta z(\cdot,\gamma) ||_{\Omega_R}$ is independent of k. However $|| |x| \Delta z(\cdot,\gamma) ||_{\Omega_R}$ tends to infinity for $\gamma \to \infty$, and I must choose $\gamma(k)$ such that $\gamma(k) \to +\infty$ for $|k| \to +\infty$, to get the other terms small. But I am free to choose for $\gamma(k)$ a function which tends to infinity sufficiently slowly, such that $|| |x| \Delta z(\cdot,\gamma(k)) ||_{\Omega_R} = o(|k|)$ is satisfied, which implies

$$|k|^{-1} || |x| \Delta z(\cdot,\gamma(k)) ||_{\Omega_R} \to 0$$

for $|k| \to \infty$. So I have $||u-u_\chi||_{\Omega_R}$ 0 and hence $||u-u_{GO}||_{\Omega_R} \to 0$, for $|k| \to \infty$, and I am finished.

Note that the essential point of the inequality (1) is that $||u||^2_{\Omega_R}$ is multiplied by k^2, at least, as far as this proof is concerned.

Dr. H. D. Alber,
Institut fur Angewandte Mathematik der Universitat Bonn,
Abteilung fur Mathematische Methoden der Physik,
Wegelerstr. 10, 53 Bonn, West Germany.

Asymptotic expansions for eigenvalues and scattering frequencies in stiff problems and singular perturbations

E. SANCHEZ-PALENCIA

1. INTRODUCTION

The purpose of this communication is to show how formal asymptotic expansions of the boundary layer or related kinds may furnish information in spectral perturbation problems for which a rigorous theory is not available.

We consider two independent problems; for each one, only the first term with respect to a small parameter ε is studied, and this in the case where the unperturbed eigenvalue is simple.

The first problem is the Helmholtz resonator where a small hole binds a closed cavity with an outer unbounded region. It is known [1] that some scattering frequencies of the problem with hole converge to the eigenfrequencies of the closed cavity. We study the first term of the perturbation of the scattering frequency as a function of the radius of the hole. The singularity of the scattering function near the hole is described in terms of the Lions-Magenes theory [6] of transposition solutions of elliptic problems in Sobolev spaces of negative order, which furnishes existence and uniqueness theorems for the studied terms. This method was formerly used in [9]. See also [7], [10] for other problems with small holes.

The second problem concerns the perturbations produced by a slight viscosity in the acoustic eigenvibrations of a gas in a closed cavity. Mathematically we have a singular perturbation for a nonselfadjoint elliptic system without compactness properties.

The spectrum near the origin may be studied via a stiff perturbation and Kato's method ([4], [3], [5]). For the other regions, we use an expansion with boundary layers of the type Vishik-Lyusternik [12] where the tangential and normal components of the velocity vector play different roles. Only the two-dimensional problem is considered.

The classical matching procedures for expansions in different regions [2], [11] are used.

2. THE HELMHOLTZ RESONATOR

We consider the scattering frequencies of the Helmholtz resonator shown in fig. 1. In the plane $\mathbb{R}^2$ of the variables y_1, y_2 we consider a bounded domain T^1 (the unit hole) containing the origin. Let $T^\varepsilon = \varepsilon\, T^1$ be its homothetic with ratio ε (parameter which $\to 0$). In the space $\mathbb{R}^3$ of the variables x_1, x_2, x_3 we consider the domain

$$\Omega^\varepsilon = \Omega^+ \cup \Omega^- \cup \bar{T}^\varepsilon \tag{2.1}$$

where Ω^+, Ω^- are respectively inner and outer domains having a neighbourhood of the origin of the $x_3 = 0$ plane as common boundary (see fig. 1).

It is known ([1], [8] chap. 17) that for small ε, the Neumann problem in Ω^ε has a scattering frequency $\sqrt{\lambda^\varepsilon}$ which converges as $\varepsilon \to 0$ to a given eigenfrequency $\sqrt{\lambda^o}$ of the interior Neumann problem in Ω^+.

$$(\Delta + \lambda^\varepsilon) u^\varepsilon = 0 \quad \text{in } \Omega^\varepsilon \tag{2.2}$$

$$\partial u^\varepsilon / \partial n = 0 \quad \text{in } \partial\Omega^\varepsilon \tag{2.3}$$

$$u^\varepsilon \text{ is for large r a convolution of } \quad \frac{\exp(i\sqrt[3]{\lambda}\, r)}{r} \; ; \; \mathrm{Im}\sqrt{\lambda^\varepsilon} < 0 \tag{2.4}$$

We postulate the expansions

$$\lambda^\varepsilon = \lambda^o + \varepsilon\lambda^1 + \ldots \tag{2.5}$$

$$u^\varepsilon(x) = u^o(x) + \varepsilon u^1(x) + \ldots \quad \text{in } \Omega^\pm \tag{2.6}$$

and, in the vicinity of the hole an "inner expansion"

$$u^\varepsilon(x) = v^o(y) + \varepsilon v'(y) + \ldots \; ; \; y = \frac{x}{\varepsilon} \; ; \; y \in \mathbb{R}^3_{T^1} \tag{2.7}$$

where $\mathbb{R}^3_{T^1}$ is (fig. 2) the union of the unit hole and the two half-spaces $y_3 \gtrless 0$. In (2.5), (2.6), λ^o, u^o are of course the given eigenvalue and eigenfrequency of Ω^+; if the indexes $\pm$ denote the restriction to $\Omega^\pm$, we have

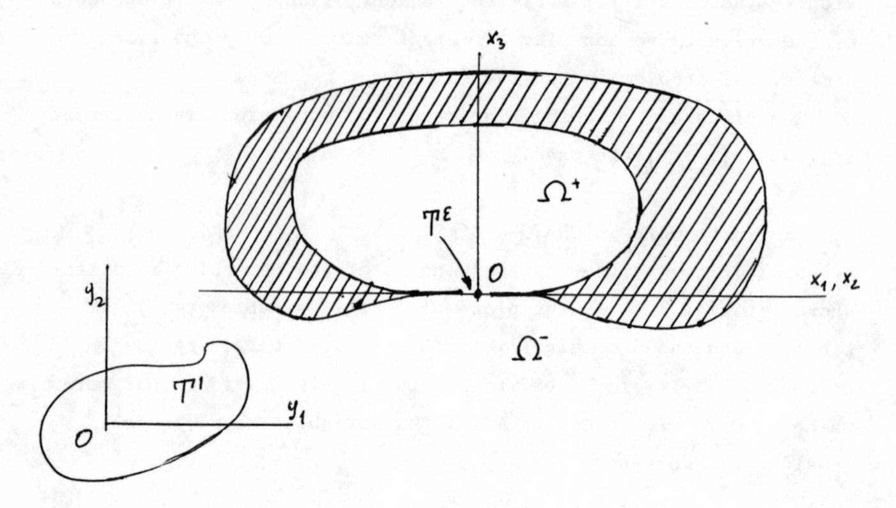

Figure 1

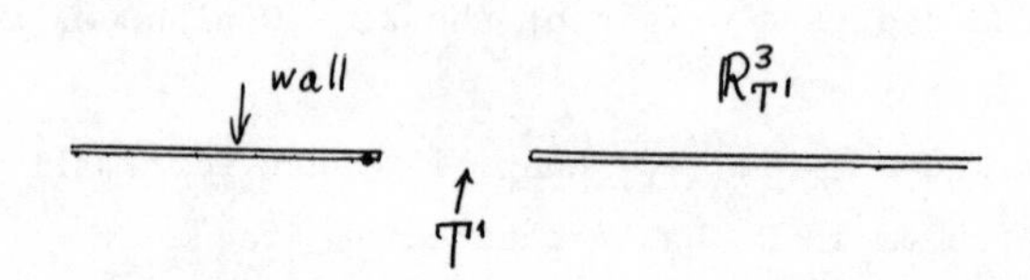

Figure 2

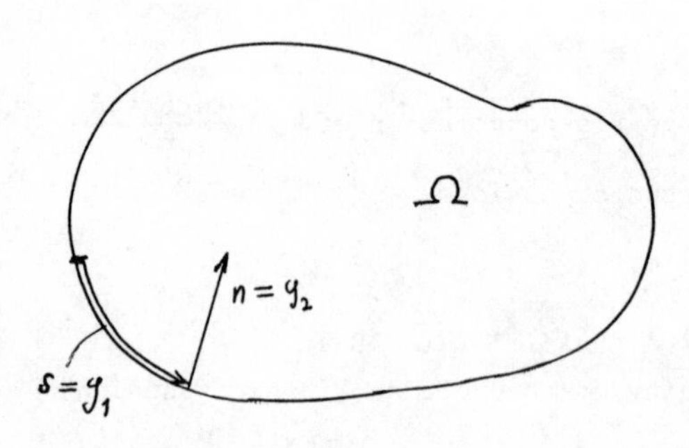

Figure 3

$$\begin{cases} u^{o+} \text{ is the solution of} \\ \text{and } u^{o-} \equiv 0 \end{cases} \begin{cases} (\Delta+\lambda^o)u^{o+} = 0 \\ \partial u^o/\partial n = 0 \\ \int_{\Omega^+} |u^{o+}|^2 dx = 1 \end{cases} \tag{2.8}$$

Study of $v^o(y)$: by setting (2.7) into (2.2), (2.3)

$$y = \frac{x}{\varepsilon} \; ; \quad \frac{\partial}{\partial x} = \frac{1}{\varepsilon}\frac{\partial}{\partial y} \tag{2.9}$$

we have

$$\Delta_y v^o = 0 \quad \text{in } \mathbb{R}^3_{T^1} \tag{2.10}$$

$$\left.\frac{\partial v^o}{\partial n}\right|_{\text{wall}} = 0 \tag{2.11}$$

Moreover, the expansions (2.6), (2.7) must match; the classical matching principle ([2], [11]) gives

$$\begin{cases} v^o(y) \xrightarrow[|y|\to\infty, y_3>0]{} u^{o+}(0) \\ v^o(y) \xrightarrow[|y|\to\infty, y_3<0]{} u^{o-}(0) \end{cases} \tag{2.12}$$

This problem (2.10) - (2.12) is well posed. An elegant proof may be obtained from the Babinet's complementary property as follows. We write:

$$v^o(y) = \frac{1}{2} u^{o+}(0) + \frac{1}{2} u^{o+}(0) V(y) \tag{2.13}$$

where $V(y)$ is a normalized function satisfying (2.10), (2.11) and taking the limits ± 1 as $|y| \to \infty$ with $y_3 \gtrless 0$ instead of (2.12). Let $W(y)$ be the solution of the capacity problem for T^1 in $\mathbb{R}^3$:

$$\begin{cases} -\Delta_y W = 0 \\ W\big|_{T^1} = 1 \\ W \xrightarrow[|y|\to\infty]{} 0 \end{cases} \tag{2.14}$$

it is easy to see that the function

$$V(y_1,y_2,y_3) + \begin{cases} 1-W(y_1,y_2,y_3) & \text{if } y_3 > 0 \\ -1+W(y_1,y_2,-y_3) & \text{if } y_3 < 0 \end{cases} \tag{2.15}$$

is harmonic in $\mathbb{R}^3_{T^1}$ and is in fact the function we searched for. Moreover, if the capacity of T^1 is defined by

$$\mathcal{C} = \frac{-1}{4\pi} \int_\Sigma \frac{\partial W}{\partial n}\, ds$$

where Σ is a any closed surface enclosing T^1, by taking as Σ the surface T^1 (two faces) itself, we obtain by the symmetry of W: (T^{1+} is the face $y_3 > 0$ of T^1):

$$\int_{T^{1+}} \frac{\partial V}{\partial y_3} dy_1 dy_2 = 2\pi\mathcal{C} \tag{2.16}$$

Study of λ^1 and u^{1+}: equation (2.2) to order ε gives

$$(\Delta+\lambda^o)u^1 = -\lambda^1 u^o \qquad x \in \Omega^{\pm} \tag{2.17}$$

We now search for the boundary values of u^1 in $\partial\Omega^+$ and $\partial\Omega^-$ (the two regions will be studied separately). Since $\partial\Omega^+$ is formed by T^ε and a part of $\partial\Omega^\varepsilon$, on T^ε we may use (2.7), (2.13). On $\partial\Omega^\varepsilon$ we use (2.3) which implies

$$\frac{\partial u^\varepsilon}{\partial n} = \cancel{\frac{\partial u^o}{\partial n}} + \varepsilon \frac{\partial u}{\partial n} + \ldots = -\frac{1}{\varepsilon}\frac{\partial v^o}{\partial y_3}(\frac{x}{\varepsilon}) + \ldots =$$

$$= \varepsilon\left[\mp \frac{1}{\varepsilon^2}\frac{u^{o+}(0)}{2}\frac{V}{\partial y_3}(\frac{x}{\varepsilon}) + \ldots\right] \quad \text{for } \Omega^{\pm} . \tag{2.18}$$

We divide by ε and by using the fact that

$$\frac{1}{\varepsilon^2}\frac{V}{\partial y_3}(\frac{x}{\varepsilon})\bigg|_{x_3=0} \to 2\pi\,\delta \text{ in } \mathcal{D}' \tag{2.19}$$

in the sense of distributions on x_1, x_2 (see Remark 2.1 below for details) we have

$$\frac{\partial u^{1\pm}}{\partial n} = \pm\pi\mathcal{C}u^{o+}(0)\delta \in H^{-s}(x_1,x_2) \quad \text{for } s > 1 \tag{2.20}$$

We consider (2.17), (2.20) in Ω^+ (the unknown is then u^{1+}). Because λ^o is an eigenvalue of the Neumann problem in Ω^+, if $\lambda^1 u^o$ is considered as known we will have a compatibility condition for the existence of u^1. It is obtained by multiplying (2.17) by u^{o+} and bearing in mind that u^{o+} is an eigenfunction and by using (2.20)

$$\int_{\partial\Omega+} \frac{\partial u^1}{\partial n} u^o ds = -\lambda^1 \int_{\Omega+} |u^o|^2 dx$$

$$\Leftrightarrow \quad \lambda^1 = \pi\mathcal{C}|u^{o+}(0)|^2 \tag{2.21}$$

and λ^1 is obtained. If this compatibility condition is fulfilled, equation (2.17) gives u^{1+} up to an eigenvector

$$u^{1+} = \hat{u}^{1+} + \alpha u^{o+} = \hat{u}^{1+}$$

where $\hat{u}^{1+}$ is orthogonal to u^{o+} and $\alpha = 0$ is obtained from a normalization condition, for instance

$$\int_{\Omega^+} u^\varepsilon u^o dx = 0 \Rightarrow \int_{\Omega^+} u\, u^0 dx = 0$$

Consequently λ^1, u^{1+} are well determined, and $u^{1+} \in H^{-s+3/2}$, $s > 1$.

Study of u^{1-}: we use (2.17), (2.20) and (2.4) to obtain

$$(\Delta+\lambda^o)u^{1-} = 0 \quad \text{in } \Omega^- \tag{2.22}$$

$$\frac{\partial u^{1-}}{\partial n} = \pi \mathcal{C} u^{o+}(0)\delta \quad \text{on } \partial\Omega^- \tag{2.23}$$

u^{1-} is for large r a convolution of $\dfrac{\exp(i\sqrt{\lambda^o}r)}{r}$ (2.24)

and it is easily seen that

$$u^{1-} = -\frac{\mathcal{C}u^{o+}(0)}{2}\frac{e^{i\lambda^o r}}{r} + \left[\text{regular function at the origin}\right]$$

Remark 2.1

We may write (2.19) in a more precise manner by using an expansion in moments. Let $\phi(y)$ be any function with compact support in $\mathbb{R}^2$ (the result is to be applied to $\partial V/\partial y_3$); we define the moments

$$\beta^o = \int_{\mathbb{R}^2} \phi(y)dy$$

$$\beta_i^o = \int_{\mathbb{R}^2} \phi(y)y_i dy$$

$\phi(x/\varepsilon)$ has a support shrinking to the origin; by acting on a test function $\theta(x)$ we see that only the values of x near the origin have an influence on the result. We then have

$$<\theta(\tfrac{x}{\varepsilon}),\theta(x)> = \int_{\mathbb{R}^2} \theta(\tfrac{x}{\varepsilon})\left[\theta(0) + \frac{\partial\theta(0)}{\partial x_i}x_i + \ldots\right]dx =$$

$$= (\text{since } x = \varepsilon y;\ dx = \varepsilon^2 dy) = <\varepsilon^2\beta^o\delta - \varepsilon^3\beta_i\frac{\partial\delta}{\partial x_i} + \ldots, \theta>$$

Moreover, from

$$|<\frac{1}{\varepsilon^2}\frac{\partial V}{\partial y_3}(\tfrac{x}{\varepsilon}),\theta>| \leqslant C||\theta||_{C^o} \leqslant C^1||\theta||_{H^s} \qquad (s > 1)$$

we see that (2.19) holds in H^{-s}, $s > 1$.

3. EIGENFREQUENCIES OF A SLIGHTLY VISCOUS GAS IN A BOUNDED REGION

If $\varepsilon\mu, \varepsilon\lambda$ are the viscosity coefficients of a compressible, barotropic gas, the eigenvalue problem amounts to find the spectral parameter z^ε and the eigenvector $\underline{u}^\varepsilon$ (see [4] for details) satisfying

$$\begin{cases} -\mu\varepsilon z^\varepsilon \Delta \underline{u}^\varepsilon + [1-\varepsilon(\lambda+\mu)z^\varepsilon] \text{grad div } \underline{u}^3 = (z^\varepsilon)^2 \underline{u}^\varepsilon \text{ in } \Omega \\ \underline{u}^\varepsilon = 0 \quad \text{on } \partial\Omega \end{cases} \tag{3.1}$$

where Ω is a bounded domain of R^2 (see fig. 3).

According to [12], we postulate the expansions:

$$\begin{cases} \underline{u}^\varepsilon = \underline{u}^o(x) + \sqrt{\varepsilon}\underline{u}^1(x) + \varepsilon\underline{u}^2(x) + \ldots \\ z^\varepsilon = z^o + \sqrt{\varepsilon}z^1 + \varepsilon z^2 + \ldots \end{cases} \tag{3.2}$$

and (3.1) gives to orders ε and $\sqrt{\varepsilon}$ respectively

$$\text{grad div } \underline{u}^o = (z^o)^2 \underline{u}^o \tag{3.3}$$

$$\text{grad div } \underline{u}^1 - (z^o)^2 \underline{u}^1 = 2z^o z^1 \underline{u}^o \tag{3.4}$$

Moreover, in the vicinity of $\partial\Omega$ we consider the curvilinear coordinates $y_1 = s$, $y_2 = n$ (see fig. 3), and the boundary layer expansion

$$u^\varepsilon \Big|_{B.L} = v^o(N,s) + \sqrt{\varepsilon} v (N,s) + \varepsilon \ldots; \quad N = \frac{n}{\sqrt{\varepsilon}} \tag{3.5}$$

By setting (3.5) into (3.1) we obtain to order ε^{-1} for the normal component v_N^o

$$\frac{\partial^2 v_N^o}{\partial N^2} = 0 \Rightarrow v_N^o \equiv 0 \Rightarrow \boxed{\underline{u}^o \cdot \underline{n}\Big|_{\partial\Omega} = 0}$$

where the arrows are consequence of $(3.1)_2$ and the standard matching with the outer expansion (3.2). We then have the boundary condition for $\underline{u}^o$. With (3.3) and $z^o \neq 0$ (see [3] for the case $z^o = 0$) we have the classical inviscid eigenoscillations $z^o = i\omega$ and $\underline{u}^o$ is the associated eigenfunction.

Moreover, (3.1), (3.5) gives, to order $\varepsilon^{-\frac{1}{2}}$ in the normal component and ε^o in the tangential component the following system for v_s^o, v_N^1

$$\begin{cases} \dfrac{\partial}{\partial N}\left(\dfrac{\partial v_s^o}{\partial s} + \dfrac{\partial v_N^1}{\partial N}\right) \quad 0 \\ -\mu \dfrac{\partial^2 v_s^o}{\partial N^2} + \dfrac{1}{i\omega}\dfrac{\partial}{\partial s}\left(\dfrac{\partial v^o}{\partial s} + \dfrac{\partial v_N^1}{\partial N}\right) \quad i\omega v_s^o \end{cases} \tag{3.7}$$

This system is easily integrated; by using the boundary condition on N = 0 and the matching

$$\lim_{N \to +\infty} v_s^o = u_s^o\Big|_{\partial\Omega}$$

We obtain

$$\begin{cases} v_s^o = u_s^o\Big|_{\partial\Omega} (1-e^{aN}); \quad a \quad -\sqrt{\dfrac{\omega}{2\mu}}\,(1+i) \\ v_N^o = \dfrac{1}{a}\,(1-e^{aN})\,\dfrac{d}{ds}\,(u_s^o\Big|_{\partial\Omega}) \end{cases} \tag{3.8}$$

But if the boundary layer term v_N^1 is known, the matching condition

$$\underline{u}^1 \cdot \underline{n}\Big|_{\partial\Omega} = \lim_{N\to\infty} v_N^1(N) \equiv \frac{1}{a}\frac{d}{ds}\,(u_s^o\Big|_{\partial\Omega}) \tag{3.9}$$

is the boundary condition to be fulfilled by $\underline{u}^1(x)$ because $\underline{u}^o$ is known. Consequently we must solve (3.4), (3.9); if $\underline{u}^1$ is the unknown, we have the compatibility condition obtained by multiplying (3.4) by the eigenfunction $\underline{u}^o$

$$2z^o z^1 \int_\Omega |u^o|^2 dx + \frac{1}{a}\int_{\partial\Omega} \frac{d}{ds}\,(u_s^o\Big|_{\partial\Omega})\,u_s^o ds = 0\ . \tag{3.10}$$

By using a normalization condition

$$\int_\Omega \underline{u}^\varepsilon\, \underline{u}^o\, dx = 1 \Rightarrow \int_\Omega |u^o|^2 dx = 1; \int_\Omega \underline{u}^1 \cdot \underline{u}^o\, dx = 0 \tag{3.11}$$

we obtain z^1 from (3.10). The compatibility condition is then satisfied and $\underline{u}^1$ is obtained in a unique way from (3.4), (3.9) with the orthogonality condition (3.11).

REFERENCES

1. J. T. Beale: Scattering frequencies of resonators, Comm. Pure Appl. Math. 26, 549-563, 1973

2. W. Eckhaus: Asymptotic Analysis of Singular Perturbations, North-Holland, Amsterdam, 1979.

3. G. Geymonat, M. Lobo-Hidalgo and E. Sanchez-Palencia: Spectral properties of certain stiff problems in elasticity and acoustics, Math. Meth. Appl. Sci., (to appear).

4. G. Geymonat and E. Sanchez-Palencia: On the vanishing viscosity limit for acoustic phenomena in a bounded region, Arch. Rat. Mech. Anal. 75, 257-268, 1981.

5. T. Kato: Perturbation theory for linear operators, Springer, New York, 1966.

6. J. Lions and E. Magenes: Problèmes aux limites et applications I, Dunod, Paris, 1968.

7. J. Sanchez-Hubert and E. Sanchez-Palencia: Acoustic fluid flow through holes and permeability of perforated walls, Jour. Math. Anal. Appl. (to appear).

8. E. Sanchez-Palencia: Non homogeneous media and vibration theory, Springer, Berlin, 1980 (= Lect. Notes Phys. 127).

9. E. Sanchez-Palencia: Boundary value problems in domains containing perforated walls, Collège de France, Seminar on Nonlinear, P.D.E., (1980-81) ed. by H. Brézis and J. L. Lions, Pitman, London (to appear).

10. E. O. Tuck: Matching problems involving flow through small holes, in Advan. Appl. Mech., 15, 89-158, Academic Press, New York, 1975.

11. M. Van Dyke: Perturbation Methods in Fluid Mechanics, Academic Press, New York, 1964.

12. M. I. Vishik and L. A. Lyusternik: Regular degeneration and boundary layer for linear differential equations with a small parameter, Usp. Mat. Nauk, 12, 3-122, 1957 (= Amer. Math. Soc. Transl. ser. 2, 20, 239-364, 1962).

Professor E. Sanchez-Palencia
Laboratoire de Mécanique Theorique
Université Paris VI
4 Place Jussieu, 75230 Paris, France

Assessment of the degree of instability in parametric resonances

E. JOSS

SYNOPSIS

This paper presents a method of calculating the real parts of the characteristic exponents which determine the degree of instability in the neighbourhood of parametric resonances. The limits of stability and the damping necessary to assure stability can also be computed by this method.

The method is quite feasible for systems with many degrees of freedom using digital computation.

Results are compared with a much slower method involving repeated numerical integration of the equations of motion over one period of the parameter variation starting from different initial conditions in order to obtain the transition matrix. Agreement between the two methods is very good even for quite large amplitudes of parameter variation.

Agreement is also good with the results of computer simulation and experimental measurement.

1. INTRODUCTION

Parametric resonances may occur in systems where the parameters are subject to periodic variation. In rotating machinery this can occur if the shafts or rotors are not solids of revolution. If the supports are flexible and anisotropic, instability can occur when the shaft speed is equal to any of the natural frequencies (simple resonance), or to the arithmetic mean of pairs of natural frequncies (combination resonance). Although the basic Floquet theory suggests that instability may also occur when the shaft speed is equal to the semi-difference of two natural frequencies, this is not found to be so either experimentally or computationally for the simple single mass system analysed. These problems have been discussed by many authors, using various methods, for example Foote, Poritsky and Slade (1), Black and McTernan (2), Yamamoto, Ota and Kono (3), Parszewski and Krodkiewski (4), Krodkiewski (5).

Literature on the mathematical theory of differential equations with periodic coefficients is extensive (the bibliography in the survey paper of Starzinskii (6) contains 146 entries stretching well back into the 19th century), but only that of Floquet (7), will be mentioned in particular.

The perturbation-variation method described by Hsu (8,9,10) is attractive for engineering applications because of the insight it gives on the nature of the motions near parametric resonances and has been used in some of the papers mentioned above.

This paper presents a method, based on the work of Malkin (11) of calculating the degree of instability at parametric resonances, the limits of stability and also the amount of damping required to assure stability. An error in Malkin's work is pointed out.

The method has been applied to two problems in gyrodynamics and one in rotordynamics and the results are compared with computer simulations and experiment.

2. OUTLINE OF CLASSICAL THEORY

The equations to be dealt with are of the form:

$$\dot{x}(\tau)=(a+\mu f)x(\tau) \qquad (1)$$

where $x(\tau)$ is an nx1 vector describing the state of the system at time τ, a is an nxn constant matrix, f is an nxn matrix of periodic coefficients, period T, and μ is a "small" quantity representing the amplitude of variation of the parameters.

The classical theory of Floquet (7) states that the vector $x(\tau) = \lceil x_1(\tau), x_2(\tau) \ldots x_n(\tau) \rfloor$ is related to the vector one period later by the equation:

$$X(\tau+T) = CX(\tau) \tag{2}$$

where C is a constant nxn matrix.

The eigenvalues of the transition matrix C are termed the characteristic multipliers, $k_1, k_2, \ldots, k_n$ say. The characteristic exponents $\alpha_1, \alpha_2, \ldots, \alpha_n$ are related to the characteristic multipliers as follows:

$$k_i = e^{\alpha_i T} \quad i = 1, 2, \ldots, n$$

If any of the characteristic multipliers lies outside the unit circle of an Argand diagram, or if any characteristic exponent has a positive real part, the system is unstable.

Expresssions of the form:

$$x_i = e^{\alpha T} \phi_i(\tau) \tag{3}$$

are solutions of the equations (1), where the $\phi_i(\tau)$ are periodic, period T. In an unstable zone the real part of α is positive and at the stability limits on either side of the zone $Re(\alpha)=0$

3. OUTLINE OF MALKIN'S METHOD

Malkin (11) gives several methods of dealing with this problem. but the method given in his Chapter V, Sec. 7 seems to be the most useful.

The equations (1) are first non-dimensionalised by the change of variable $t = \frac{1}{2}\omega_o \tau$ where $\omega_o = \frac{2\pi}{T}$ is the parameter variation frequency.

In the case of rotating machinery, ω_o is usually twice the shaft angular velocity ω_s.

The equations then become

$$\dot{x}(t) = \frac{1}{\frac{1}{2}\omega_o} \dot{x}(\tau) = \lambda (a+\mu f) x(t) \tag{4}$$

where $\lambda=2/\omega_o$ and the coefficients f now have period π.

In the equation (1) or (4) the coefficients a and f are split up so that all the damping terms and periodic terms are contained in the matrix f, leaving matrix a to represent a system having eigenvalues in conjugate imaginary pairs.

To find the limits of stability the characteristic exponents are calculated for different values of the parameter variation frequency near resonance. Variation of this frequency is represented by the quantity σ where $\lambda=\lambda_o+\mu\sigma$ and λ_o is taken as one of the values producing parametric resonance.

CONDITIONS FOR PARAMETRIC RESONANCE

If $\omega_1, \omega_2, \ldots, \omega_m$ where m=n/2 are the natural frequencies of the original system for $\mu=0$, the eigenvalues of the matrix are $\pm i\omega_1, \pm i\omega_2, \ldots \pm i\omega_m$. The condition for parametric resonance is that the difference between any two eigenvalues shall be an integer multiple of ω_o

i.e. $i(\omega_j \pm \omega_k)=Ni\omega_o$ $(N=\pm1,\pm2,\ldots)(j,k=1,2,\ldots,m)$

or, dividing by i

$$\lambda(\omega_j \pm \omega_k)=2N \tag{5}$$

For simple parametric resonance k=j and $2\omega_j=\omega_o$, i.e. $\omega_s=\omega_j$. For $k\neq j$ combination resonance occurs if the above condition (5) is satisfied.

CALCULATION OF CHARACTERISTIC EXPONENTS

The characteristic exponent of equations (4) is expressed in the form $\alpha=\lambda_o i\omega_j+\mu\bar{a}$ where λ_o satisfies relation (5) above. The sign of the real part $\mu\bar{a}$ determines whether or not the system is stable. Making the change of variable $x=e^{\alpha t}y$ the following system of equations is obtained which has two periodic solutions, one a vector of constants (i.e. periodic, arbitrary period) and one a vector of functions, period Π.

$$\dot{y}=(\lambda_o+\mu\sigma)(a+\mu f)y-(\lambda_o i\omega_j+\mu\bar{a})I_n y \tag{6}$$

where I_n is the n×n identity matrix.

The relationship between the pth and (p - 1)th approximations to the solution of equations (6) is as follows:

$$\dot{y}^{(p)}=\lambda_o(a-i\omega_j I_n)y^{(p)}+\mu(\sigma a+\lambda_o f+\mu\sigma f)y^{(p-1)}-\mu\bar{a}^{(p)}I_n y^{(p-1)} \tag{7}$$

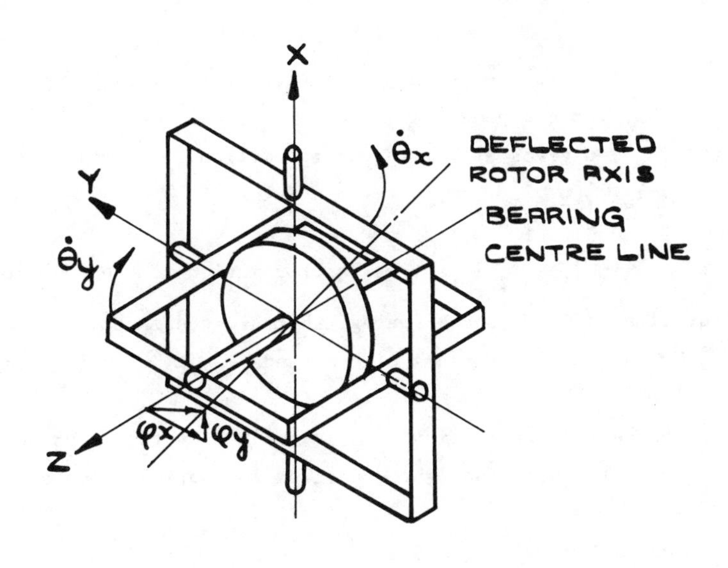

FIG. 1

THREE AXIS GYROSCOPE

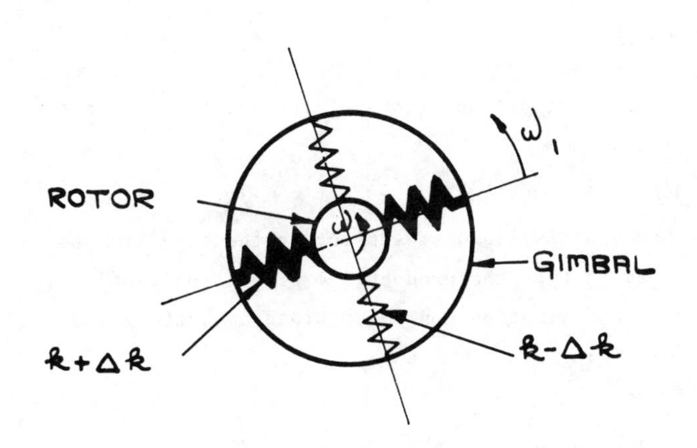

FIG. 2

DIAGRAMATIC REPRESENTATION OF STIFFNESS VARIATION IN ROTOR BEARINGS

where $\mu\,\bar{a}^{(p)}$ is the pth approximation to the real part of the characteristic exponent α.

Putting $\mu=0$ gives as a first approximation to equation (6)

$$\dot{y}^{(o)}=\lambda_o(a-i\omega_j I_n)y^{(o)} \tag{8}$$

and $y^{(o)}=M_o\Phi_1+N_o\Phi_2$ is a first approximation to the solution of equations (6), M_o and N_o being arbitrary constants.

The vector Φ_1 consists of constants, Φ_2 of periodic terms, period Π.

The next approximation $y^{(1)}$ is defined by the recurrence relation (7) above and the condition for periodicity is

$$\left[\sigma A+(1+\frac{\mu\sigma}{\lambda_o})B-\bar{a}^{(1)}I_2\right]m=0 \tag{9}$$

where m is the 2×1 column vector M_o, N_o.

I_2 is a 2×2 unit matrix and A and B are 2×2 matrices defined by Malkin as follows:

$$A=\int_o^{2\pi}\Phi' a'\psi dt$$
$$B=\lambda_o\int_o^{2\pi}\Phi' f'\psi dt \tag{10}$$

where ψ_1 and ψ_2 are periodic solutions of the system conjugate to equations (6) i.e.:

$$\dot{z}+(\lambda_o+\mu\sigma)(a+\mu f)z-(\lambda_o+i\omega_j)I_n z=0 \tag{11}$$

The matrix equation (9) is also subject to the condition that the ψ's are chosen so that the product $\psi'\Phi$ gives a unit 2×2 matrix.

Again a first approximation can be obtained by putting $\mu=0$ in equation (9), giving:

$$(\sigma A^o+B^o-a^*I_2)m^*=0 \tag{12}$$

and putting the determinant $|\sigma A^o+B^o-a^*I_2|=0$ gives the quadratic equation:

$$a^{*2}-[\sigma(A_{11}{}^o+A_{22}{}^o)+B_{11}{}^o+B_{22}{}^o]a^*$$
$$+(\sigma A_{11}{}^o+B_{11}{}^o)(\sigma A_{22}{}^o+B_{22}{}^o)-(\sigma A_{21}{}^o+B_{21}{}^o)(\sigma A_{12}{}^o+B_{12}{}^o)=0 \tag{13}$$

Putting $\sigma=0$ gives a first approximation to the values of the

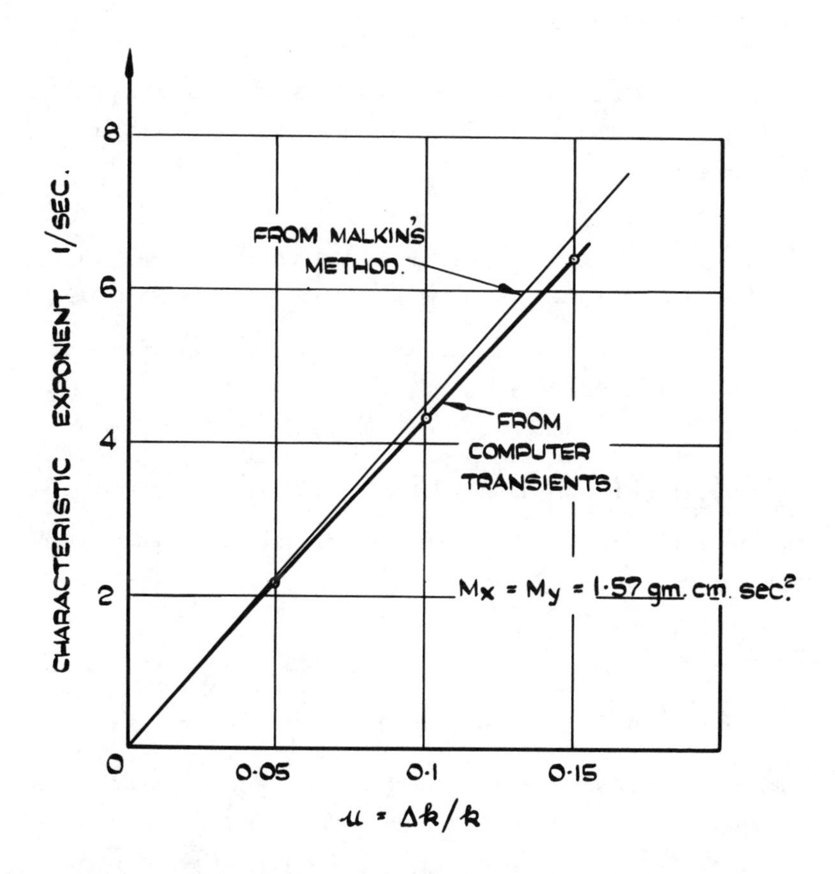

EFFECT OF USING THE FIRST APPROXIMATION IN CALCULATING THE CHARACTERISTIC EXPONENT.

FIG. 3

characteristic exponents a* at the centre of the resonance zone, and putting a*=0 gives the width of the resonance zone in terms of σ. This leads easily to an expression for the unstable zone in terms of frequency.

4. MODES OF VIBRATION ASSOCIATED WITH CHARACTERISTIC EXPONENTS

If it is assumed that the system is tuned to a combination resonance so that:

$$\omega_j - \omega_k = N\omega_o$$

as in relation (5), the first approximation $y^{(o)}$ to the solution of equations (7) is of the form $y^{(o)}=K_1+K_2e^{-2iNt}$ where K_1 and K_2 are complex constants. Returning to the original variable $x^{(o)}$ via the transformation:

$$x=e^{(\lambda_o iw_j+\mu\bar{a})t}y$$

and recalling that $\bar{a}$ has two values a_1 and a_2 (i.e. the roots of the quadratic equation (13)) the solution $x^{(o)}$ is seen to be of the form:

$$x^{(o)}=K_1e^{\mu a_1 t}e^{\lambda_o i\omega_j t}+K_2e^{\mu a_1 t}e^{(\lambda_o i\omega_j-2iN)t} \tag{14}$$
$$+K_1'e^{\mu a_2 t}e^{\lambda_o i\omega_k t}+K_2'e^{\mu a_2 t}e^{(\lambda_o i\omega_k-2iN)t}$$

Making the time scale change to real time τ by the substitutions $t=\frac{1}{2}\omega_o\tau$ and $\lambda_o=2/\omega_o$ the solution becomes:

$$x^{(o)}(\tau)=K_1e^{\frac{\mu a_1\omega_o\tau}{2}}e^{i\omega_j\tau}+K_2e^{\frac{\mu a_1\omega_o\tau}{2}}e^{(i\omega_j-iN\omega_o)\tau} \tag{15}$$
$$+K_1'e^{\frac{\mu a_2\omega_o\tau}{2}}e^{i\omega_k\tau}+K_2'e^{\frac{\mu a_2\omega_o\tau}{2}}e^{(i\omega_k-iN\omega_o)\tau}$$

Hence, there appear to be four frequencies affected by the characteristic exponents viz. ω_j, $\omega_j-N\omega_o$ associated with a_1 and ω_k, $\omega_k-N\omega_o$ associated with a_2

In fact, the frequency $\omega_j - N\omega_o = \omega_k$ because of the resonance condition (5) so that we have the higher of the two natural frequencies associated with a_1, the lower associated with both

a_1 and a_2, and a third frequency $\omega_k - N\omega_o$ associated with a_2.

Although it frequently occurs that a_1 and a_2 are equal in magnitude and opposite in sign, it is not certain that this will be true in general. However, assuming that a_1 and a_2 are at least of opposite sign, it is not certain that the unstable positive exponent will be associated with the two natural frequencies ω_j and ω_k rather than the lower natural frequency ω_k and the difference frequency $\omega_k-N\omega_o$ although intuitively it may seem more likely.

In reference (4), the authors in quoting their own earlier work, mention observing a combination resonance in which unstable vibrations are present at the two natural frequencies whose sum is equal to twice the rotational speed of the shaft.

5. APPLICATION TO GYRO WITH UNSYMMETRICAL ROTOR

Fig. 1 shows a gimbal mounted gyro with fixed references axes, OX, OY, OZ, which in the undisturbed position of the gyro correspond to the outer gimbal axis and the spin axis respectively. The rotor may deflect due to shaft or bearing compliance through the angles ϕ_x and ϕ_y relative to the inner gimbal while the gimbals themselves may be displaced through the angles θ_x and θ_y.

In this first application of Malkin's method, Φ_x and Φ_y are assumed zero and the rotor is assumed not to be a solid of revolution, so that its transverse moment of inertia about a fixed axis varies between I_1 and I_2 as it rotates.

The equations of motion then become:

$$\begin{aligned}(I+M_x)\ddot{\theta}_x+J\Omega\dot{\theta}_y+R_g\dot{\theta}_x+rI(\dot{\theta}_x\cos2\Omega t+\dot{\theta}_y\ \sin 2\omega t)&\\ +\ 2rI\Omega(-\dot{\theta}_x\sin2\Omega t+\dot{\theta}_y\ \cos2\Omega t)&=0\\ (I+M_y)\ddot{\theta}_y-J\Omega\dot{\theta}_x+R_g\dot{\theta}_y+rI(\dot{\theta}_x\sin2\Omega t-\dot{\theta}_y\ \cos2\Omega t)&\\ +\ 2rI\Omega(\dot{\theta}_x\cos2\Omega t+\dot{\theta}_y\ \sin2\Omega t)&=0\end{aligned} \qquad (16)$$

where I = mean transverse moment of inertia of rotor

$= (I_1 + I_2)/2$

$r = (I_1 - I_2)/(I_1 + I_2)$

J = moment of inertia of rotor about OZ

M_x = moment of inertia of inner and outer gimbals about OX

M_y = moment of inertia of inner gimbal about OY

Ω = angular velocity of rotor about OZ

R_g = gimbal damping coefficient

The angular velocities $\dot{\theta}_x$ and $\dot{\theta}_y$ are taken as the state variables x_1 and x_2 respectively, and after some manipulation and approximations for small μ, it can be shown that the matrices a and f are as follows:

$$a = \frac{\Omega}{(I+M_x)(I+M_y)}\begin{bmatrix} 0 & -J(I+M_y) \\ J(I+M_x) & 0 \end{bmatrix}$$

$$\mu f = \frac{r\Omega}{(I+M_x)(I+M_y)}\begin{bmatrix} I\left[2(I+M_y)-J\right]\sin 2\Omega t - R^1_g(I+M_y) & -\{J\Delta M(I+M_y) + I\left[2(I+M_y)-J\right]\}\cos 2\Omega t \\ \{J\Delta M(I+M_x) + I\left[J-2(I+M_x)\right]\}\cos 2\Omega t & I\left[J-2(I+M_x)\right]\sin 2\Omega t - R_g^1(I+M_x) \end{bmatrix} \quad (17)$$

where

$$\Delta M = \frac{I(M_x-M_y)}{(I+M_x)(I+M_y)} \quad \text{and} \quad R^1_g = \frac{Rg}{r\Omega}$$

With this two degrees of freedom system, it is relatively easy to carry out Malkin's procedure analytically. The only possible parametric resonance occurs when the natural frequency of nutation,

$\omega_n = \dfrac{J\Omega}{\sqrt{(I+M_x)(I+M_y)}}$ is equal to the spin frequency Ω of the rotor.

The analysis finally gives, for zero damping, the characteristic exponent at the centre of the resonance zone $a^*=\pm\Pi(1 + R^2)$ where

$$R = \sqrt{\frac{I+M_x}{I+M_y}} \quad \text{and the stability limits at } \omega_n = \Omega\left(1 \pm \frac{r(1+R^2)}{2}\right)$$

If the damping R_g is retained in the analysis, the damping required to keep the system stable at the centre of the resonance zone can be easily found by putting $\sigma=0$ and $A^*=0$ in equation (13).

6. COMPARISON WITH ANALOGUE COMPUTER RESULTS

The above analytical results were compared with the results of an analogue computer solution of equations (16),(reference (12)), and although the limits of stability were in excellent agreement, the characteristic exponents were in error by a factor of 2Π. Comparisons were also made with the results of one of Malkin's other methods (Chapter III, Sec. 11 of reference (11)) which gave a value of characteristic exponent agreeing with the computer results.

Further examination of Malkin's work leads back to the original derivation of the periodicity condition (9) which is to be found in Malkin's Chapter II, equation 4.13. This is quoted in Section 6 of Chapter V, equations 6.21 and 6.22, and again in Section 7 which contains the method described in this paper. The application of condition 6.22 ($\psi'\phi=1$) is evidently intended to give a unity coefficient for the exponent a_1. However, it appears that Malkin has overlooked the fact that the integration from 0 to 2Π reintroduces a coefficient 2Π and to correct for this the matrices A and B (equations (10) of this paper) should be divided by 2Π. When this is done the results of analytical work and simulations agree.

7. APPLICATION TO GYRO WITH BEARING STIFFNESS VARIATION

Referring again to Fig. 1, the displacements ϕ_x and ϕ_y are now assumed to be non zero and variation in radial stiffness of the spin axis bearings is considered, due for example, to an oversize ball. The effect is shown diagrammatically in Fig. 2 and the equations of motion for this case are given in reference (12). When put in the form of equations (1) of this paper, the coefficient matrices are as follows:

$$a = \begin{bmatrix} 0 & 1 & 0 & 0 & 0 & 0 \\ -(\frac{k}{M_x}+\frac{k}{I}) & 0 & 0 & -\frac{J\Omega}{I} & 0 & -\frac{J\Omega}{I} \\ 0 & 0 & 0 & 1 & 0 & 0 \\ 0 & \frac{J\Omega}{I} & -(\frac{k}{M_y}+\frac{k}{I}) & 0 & \frac{J\Omega}{I} & 0 \\ \frac{k}{M_x} & 0 & 0 & 0 & 0 & 0 \\ 0 & 0 & \frac{k}{M_y} & 0 & 0 & 0 \end{bmatrix}$$

$$f = \begin{bmatrix} 0 & 0 & 0 & 0 & 0 & 0 \\ -(\frac{k}{M_x}+\frac{k}{I})\cos 2\omega_1 t & 0 & -(\frac{k}{M_x}+\frac{k}{I})\sin 2\omega_1 t & 0 & 0 & 0 \\ 0 & 0 & 0 & 0 & 0 & 0 \\ -(\frac{k}{M_y}+\frac{k}{I})\sin 2\omega_1 t & 0 & (\frac{k}{M_y}+\frac{k}{I})\cos 2\omega_1 t & 0 & 0 & 0 \\ \frac{k}{M_x}\cos 2\omega_1 t & 0 & \frac{k}{M_x}\sin 2\omega_1 t & 0 & 0 & 0 \\ \frac{k}{M_y}\sin 2\omega_1 t & 0 & \frac{k}{M_y}\cos 2\omega_1 t & 0 & 0 & 0 \end{bmatrix}$$

where the variables x_1 - - x_6 correspond to ϕ_x, $\dot{\phi}_x$, ϕ_y, $\dot{\phi}_y$ $\dot{\theta}_x$, $\dot{\theta}_y$ respectively and $\mu=\Delta k/k$. ω_1 is the angular velocity of the ball bearing cage.

It is impracticable to follow the method of Sec. 3 analytically in this case, but the computation has been carried out quite readily by digital computer. The results were compared with the results of direct numerical solution of the equations on a digital computer and were found to agree within 5% for a value of $\mu=0.15$. Fig. 3 shows the comparison between the two methods for various values of μ.

The direct digital solution of the equations was carried out using initial conditions corresponding to the appropriate eigenvectors

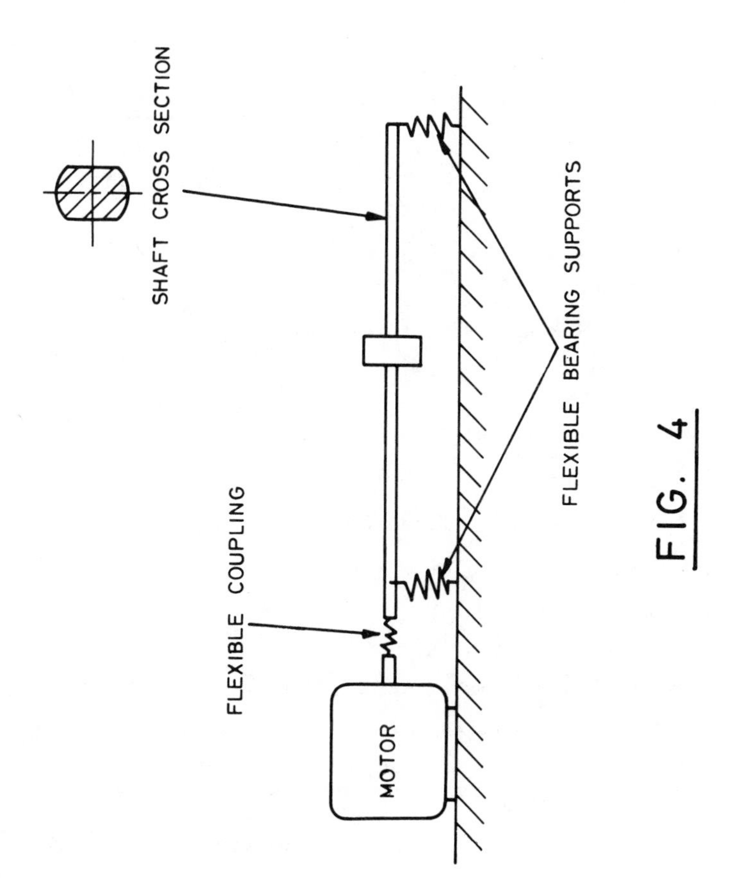
SHAFT CROSS SECTION
FLEXIBLE COUPLING
MOTOR
FLEXIBLE BEARING SUPPORTS

FIG. 4

of the matrix a, so that the motion was predisposed towards the natural frequency which was in resonance with the parameter variation frequency.

8. APPLICATION TO WHIRLING OF AN UNSYMMETRICAL SHAFT

A third mechanical system studied is shown in Fig. 4. It consists of a shaft with non-circular cross-section carrying a single balanced rotor at mid-span, the supports having greater stiffness in one direction than the other.

Analysis of the system gives the following equations of motion.

$$m\ddot{x} + c\dot{x} + Ax + By = 0$$

$$m\ddot{y} + c\dot{y} + Cx + Dy = 0$$

where:- $$A = \frac{2k_3\left[k^2-\Delta k^2+2k_4(k+\Delta k \;\; \cos 2\Omega t)\right]}{k^2-\Delta k^2+2k(k_3+k_4)+4k_3k_4-2\Delta k \; \cos 2\Omega t(k_3-k_4)}$$

$$B = C = 4\Delta k \; k_3k_4 \; \sin 2\Omega t/\text{same denom.}$$

$$D = 2k_4\left[k_2-\Delta k_2{}^2+2k_3(k-\Delta k \; \cos 2\Omega t)\right]/\text{denom}$$

and k = mean shaft stiffness, $\Delta k = (k\ max - k\ min)/2$

k_3 = bearing stiffness x direction

k_4 = bearing stiffness y direction

Ω = shaft angular velocity

c = damping coefficient

m = mass

The above method was implemented on a Honeywell 6060 computer and was found to give rapid calculation of the stability limits and characteristic exponents. All the data necessary to produce Fig. 5 was computed and written to a file in a matter of a few seconds. Alternatively a listing as in Fig. 6 can be produced with the time being determined almost entirely by the printer speed.

ACCURACY CHECK

Fig. 5, it will be seen shows values of ε up to 0.5 ($\varepsilon=\Delta k/k$) which is a very considerable degree of shaft asymmetry, corresponding to an "across flats" dimension 55% of the shaft diameter. Since the method involves assumptions of small asymmetry a different method not requiring such assumptions was employed as a check. This involved integrating the equations of motion four times, starting

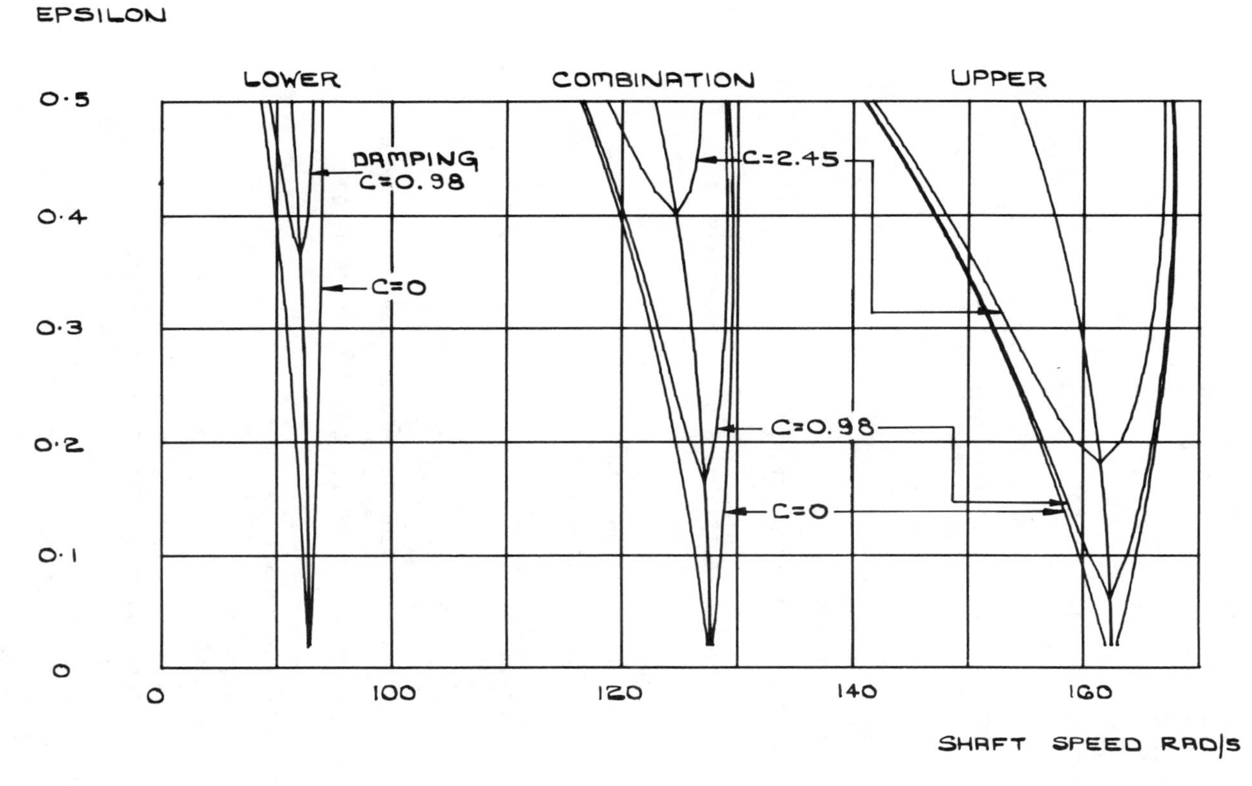

FIG. 5. STABILITY LIMITS

```
O.

parameters

mean shaft stfs.=  0.11337705E 05
bearing stfs. x=  0.14307000E 04
bearing stfs. y=  0.91215000E 04
damping         =  0.
mass            =  0.26500000E 00


EPSILON=  0.50000000E 00


LOWER COMBINATION RESONANCE
    EXP UFL AT LOCATION 021246

    EXP UFL AT LOCATION 021246

CHARACTERISTIC EXPONENTS
   0.42356298E-15
  -0.42356298E-15

    EXP UFL AT LOCATION 021245

CENTRE FREQ.           STAB.LIMS.
   0.31518809E 02  0.33172354E 02  0.29865264E 02


LOWER RESONANCE
    EXP UFL AT LOCATION 021246

    EXP UFL AT LOCATION 021246

CHARACTERISTIC EXPONENTS
   0.26900678E 01
  -0.26900678E 01

    EXP UFL AT LOCATION 021245

**THIS IS THE LAST TIME THE ABOVE MESSAGE WILL APPEAR**
CENTRE FREQ.           STAB.LIMS.
   0.91251610E 02  0.88561542E 02  0.93941678E 02


COMBINATION RESONANCE
CHARACTERISTIC EXPONENTS
   0.62249269E 01
  -0.62249269E 01

CENTRE FREQ.           STAB.LIMS.
   0.12277042E 03  0.11632962E 03  0.12921122E 03


UPPER RESONANCE
CHARACTERISTIC EXPONENTS
   0.13445986E 02
  -0.13445986E 02

CENTRE FREQ.           STAB.LIMS.
   0.15428923E 03  0.14084324E 03  0.16773521E 03
```

FIG. 6.

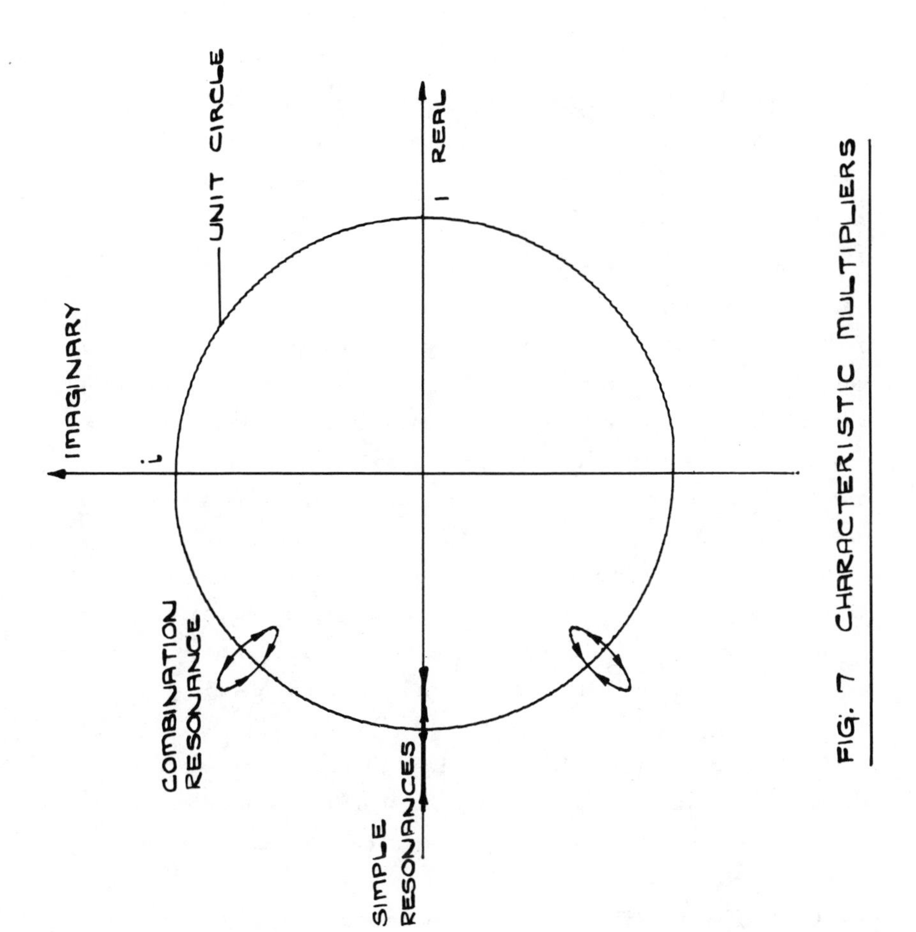

FIG. 7 CHARACTERISTIC MULTIPLIERS

from the initial conditions

$x = 1\ 0\ 0\ 0$

$\dot{x} = 0\ 1\ 0\ 0$

$y = 0\ 0\ 1\ 0$

$\dot{y} = 0\ 0\ 0\ 1$

The end condition vectors after one period together formed the transition matrix and the eigenvalues of this matrix gave the characteristic multipliers k_1 to k_4 from which the characteristic exponents α_1 to α_4 were found as $\alpha_i = \frac{1}{T} \log_e k_i$

For this system, a typical time for computing the exponents at one shaft speed is 25 s.

In general, the multipliers k are complex and Fig. 7 shows the variation of k with shaft speed. Away from the three resonances, the multipliers k travel round the unit circle ; at the simple resonances a pair of k's move inwards and outwards from the unit circle along the real axis while at the combination resonance, the multipliers form the loops in the 2nd and 3rd quadrants.

Fig. 8 shows the characteristic exponent $R_e(x)$ plotted to a base of shaft speed and it will be seen that at each resonance, positive and negative exponents appear in pairs. The loci of the exponents are very close to circles, if the horizontal and vertical scales are equal.

If damping is included, the horizontal centre lines of the exponent loci shift downwards by c/2m so that the amount of damping necessary to assure stability is easily calculated.

9. COMPARISON OF METHODS

Fig. 9 shows a comparison of the maximum positive exponents at the centre of the resonance zones as obtained by the two methods, plotted to a base of ε the degree of asymmetry.

The agreement is excellent for the upper simple and combination resonances, but not so good for the lower simple resonance.

10. EXPERIMENTAL RESULTS

The whirling shaft rig, shown in Fig. 4, has been constructed and some steady-state and transient tests have been carried out.

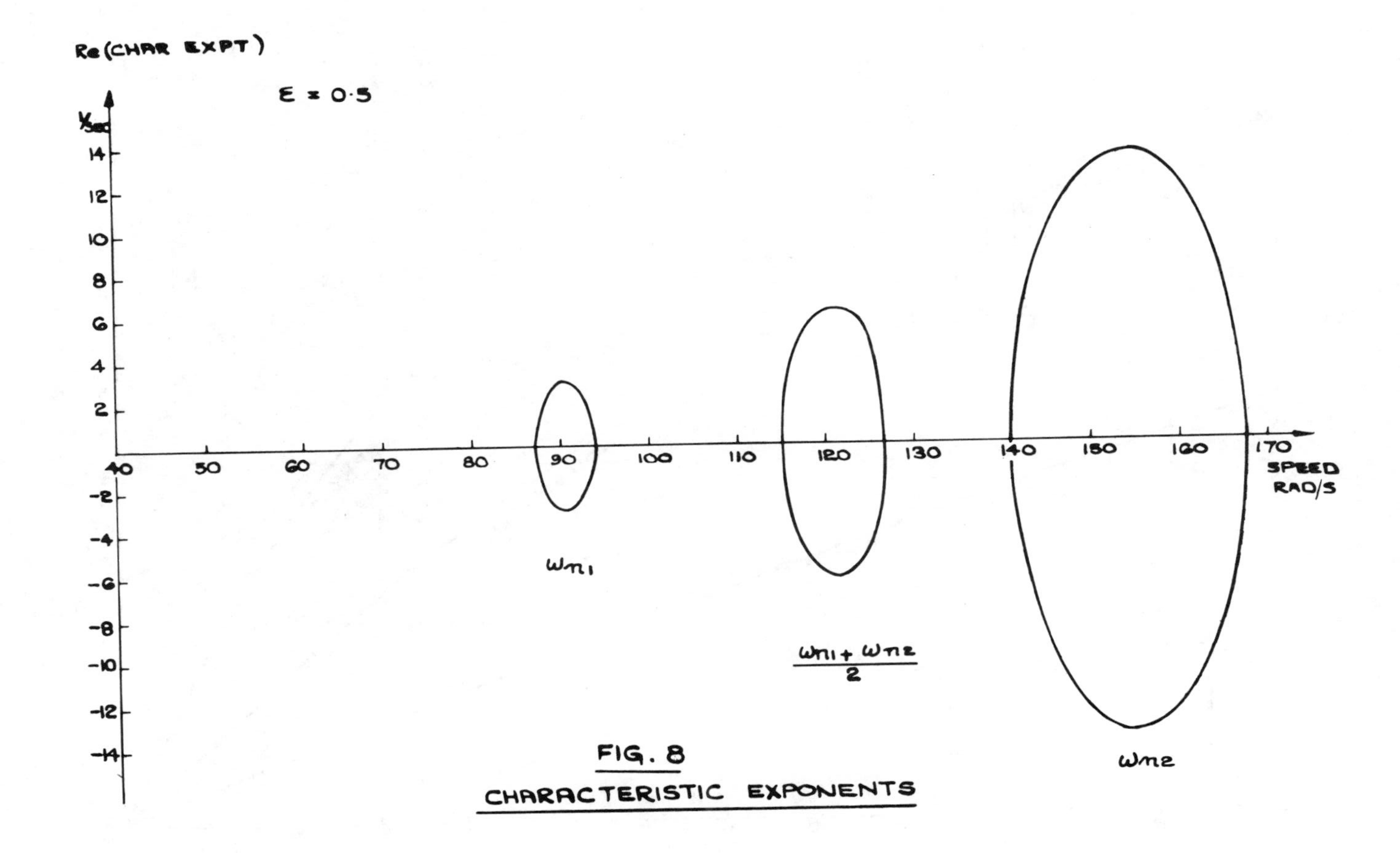

FIG. 8

CHARACTERISTIC EXPONENTS

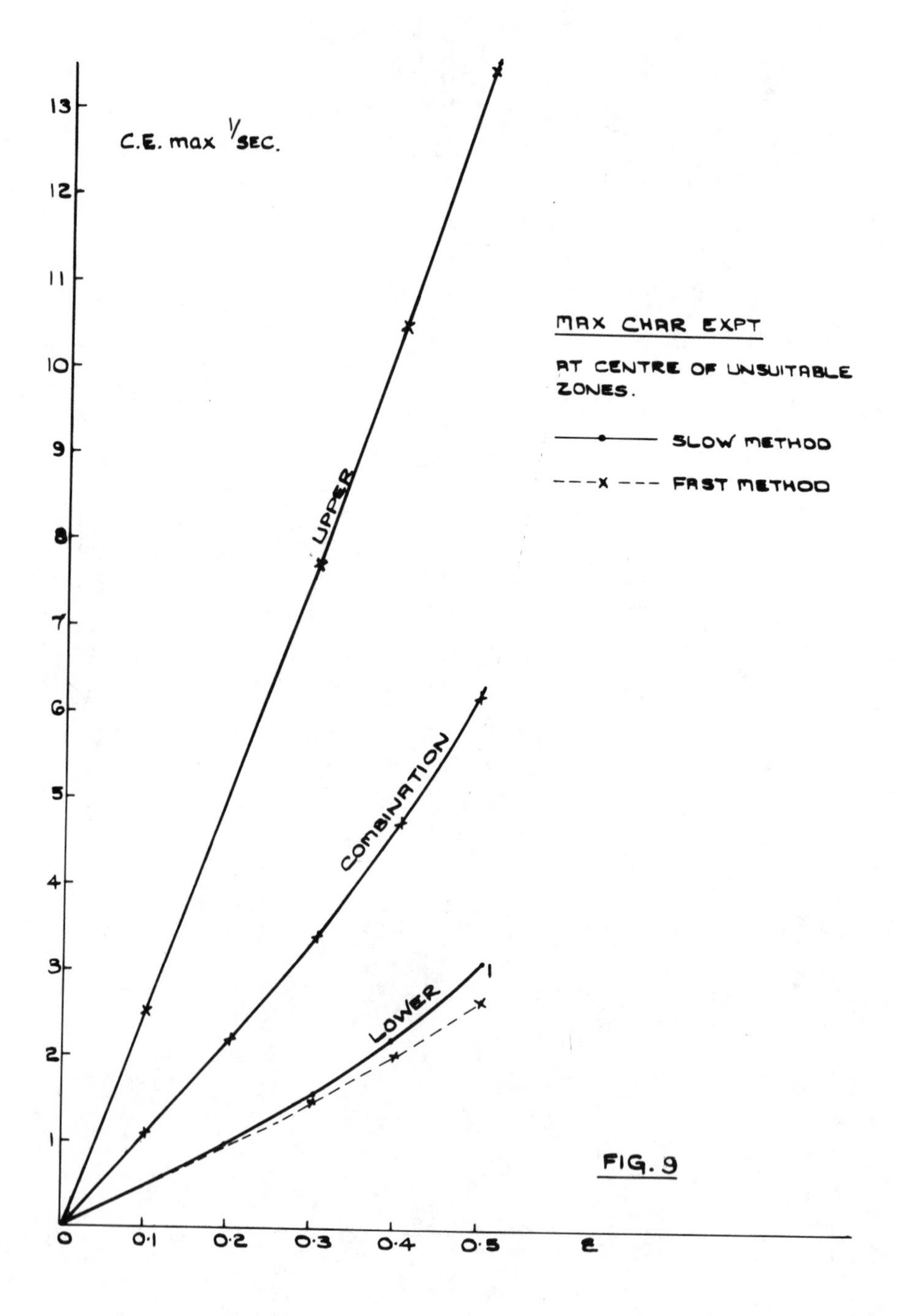

C.E. max 1/SEC.
13
12
11
10
9
8
7
6
5
4
3
2
1
0
0·1
0·2
0·3
0·4
0·5
E
UPPER
COMBINATION
LOWER
MAX CHAR EXPT
AT CENTRE OF UNSUITABLE ZONES.
SLOW METHOD
FAST METHOD

FIG. 9

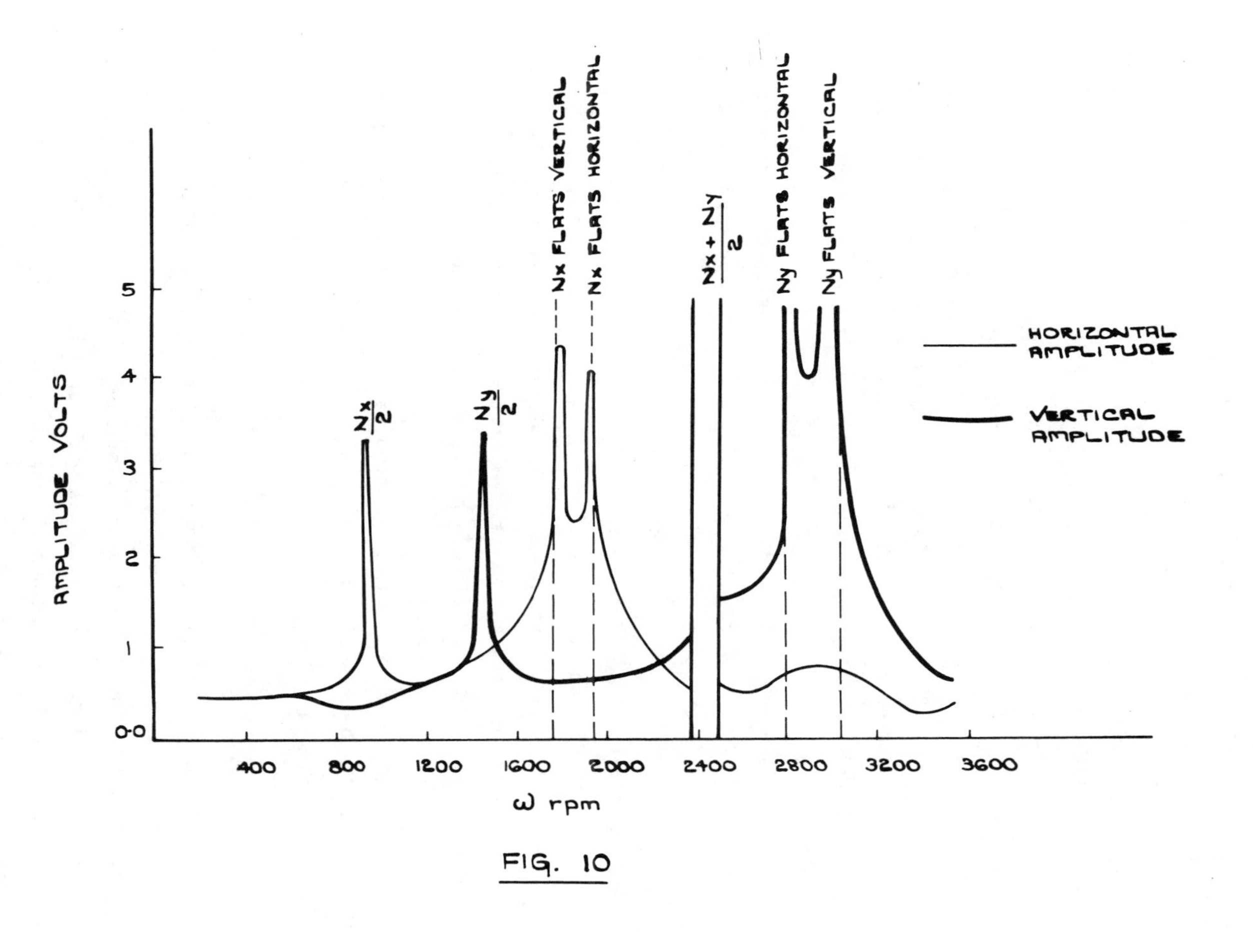

AMPLITUDE VOLTS
0·0
1
2
3
4
5
400
800
1200
1600
2000
2400
2800
3200
3600
ω rpm
$\frac{N_x}{2}$
$\frac{N_y}{2}$
Nx FLATS VERTICAL
Nx FLATS HORIZONTAL
$\frac{N_x + N_y}{2}$
Ny FLATS HORIZONTAL
Ny FLATS VERTICAL
HORIZONTAL AMPLITUDE
VERTICAL AMPLITUDE

FIG. 10

STEADY STATE TESTS

The steady-state tests are run at various steady speeds and the amplitudes of the motion of the mass in the x and y directions measured. A typical plot of amplitude against frequency is shown if Fig. 10.

In addition to the expected resonances at the two natural frequencies and their arithmetic mean, others appear when the shaft speed is equal to half the natural frequencies.

At first, the latter two resonances were assumed to be due to gravity, but running the shaft in a vertical position did not eliminate the resonance. Out of balance in the rotor or bending of the shaft will also contribute effects not taken account of in the analysis.

At high speeds, there are further modes of vibration in which the mass stays relatively still and the bearing housings vibrate, possibly a nutation-type oscillation.

TRANSIENT TESTS

These are also carried out at constant speed, but the mass is restrained by a roller-clamp device which is suddenly released and the resulting transient vibrations recorded on a storage oscilloscope. From these an attempt is made to measure the characteristic exponents in the resonant zones.

The apparatus and instrumentation will require some refinement butpreliminary results show encouraging agreement between measured and theoretical exponents.

11. CONCLUSIONS

A numerical method, using readily available eigenvalue/vector routines has been implemented which permits the rapid computation of stability limits and characteristic exponents in systems described by linear differential equations with constant coefficients. Good agreement has been obtained with a much slower numerical method based on numerical integration of the equations, computer simulations, both digital and analogue, and experimental measurements of transient whirl.

An error in the original reference (11) has been pointed out.

REFERENCES

1. Foote, W. R., Poritsky, H. and Slade, J. J. "Critical speeds of a rotor with unequal shaft flexibilities, mounted in bearings of unequal flexibility", J. Appl. Mech., Trans. Am. Soc. Mech. Engrs. 1943 65 (June) A-77.
2. Black, H. F. and McTernan, A. J. "Vibrating of a rotating asymmetric shaft supported in asymmetric bearings", J. Mech. Eng. Sci. 1968 v 10 pt 3, pp 252-261.
3. Yamamoto, T., Ota, H. and Kono, K. "On the unstable vibrations of a shaft with unsymmetrical stiffness carrying an unsymmetrical rotor", J. Appl. Mech., Trans. Am. Soc. Mech. Engrs. 1968 v 35 (June), pp 313-321.
4. Parszewski, Z. and Krodkiewski, J. "Stability and critical speeds of non-circular shafts on flexible anisotropic supports", Arch. Budowy Maszyn, Tom XXII, 1975, zeszyt 2, pp 121-137.
5. Krodkiewski, J. "Parametric vibration of a rigid body fixed on a shaft supported flexibly anisotropically", Scientific Bulletin of Lodz Technical University 1976, Nr. 223, Mechanika, z. 44.
6. Starzinskii, V. M. "On the stability of periodic motions II", American Math. Soc. Translations, Series 2, Vol. 33, pp 123-187, transl. from Bul. Inst. Politekn. Iasi Serie Nova, Tom 5 (9) 1-2 (1959), pp 51-100.
7. Floquet, G. Ann. Sci. Ecole Norm. Sup. (2) 12 (1883), pp 47-83.
8. Hsu, C. S. "On the parametric excitation of a dynamic system having multiple degrees of freedom", J. Appl. Mech., Trans. Am. Soc. Mech. Engrs. 1963 30 pp 367-372.
9. Hsu, C. S. "Further results on parametric excitation of a dynamic system", J. Appl. Mech., Trans. Am. Soc. Mech. Engrs. 1965 32, pp 373-377.
10. Hsu, C. S. and Cheng, W. H. "Steady-state response of a dynamic system under combined parametric and forcing excitations", J. Appl. Mech., Trans. Am. Soc. Mech. Engrs. 1974 41, pp 371-378.
11. Malkin, I. G. "Some problems in the theory of non-linear oscillations", Translation of Nekotorye Zadachi Teorie

Nelineinych Kolebanii" by U.S. Atomic Energy Commssion.

12. Joss, E. J. "Some aspects of gyroscope stability and dynamic response", PhD thesis, Glasgow University, 1969.

Dr. E. J. Joss,
Department of Dynamics and Control,
University of Strathclyde,
Glasgow, Scotland.

Mathematical modelling of a stretch mill

J. BROWN and R. DAVIDSON

INTRODUCTION

In aluminium sheet rolling mills, the coils of metal from the rolling process are heavily stressed and are processed to remove the deformities in a stretch-levelling machine. This stretches the material into the yield zone and levels the surface over rolls when in this yielded condition.

It will be clear that a large driving torque at the outlet from the machine is matched by corresponding braking torque at the inlet.

In many designs these torques are provided by individual motors but a significant economy in power is achieved if the braking torque can be provided mechanically within the machine.

An attractive process is to link the input and output section by gearing so that the output speed of the strip is greater than the input speed thus achieving yield.

Only epicyclic gear trains can be varied in speed ratio in the controlled manner required for this process.

The arrangement of a practical machine is further complicated by the need to build up and subsequently relax the tension in the strip in a progressive manner.

Fig. 1 shows in outline the machine for which the work described in this paper was commissioned.

Ripple marking appeared on the strip under certain operating conditions. It became clear from observation and measurement that these markings were associated with vibration.

The system is extremely complex and in order to achieve an understanding of its oscillatory behaviour, a mathematical model was deemed to be necessary.

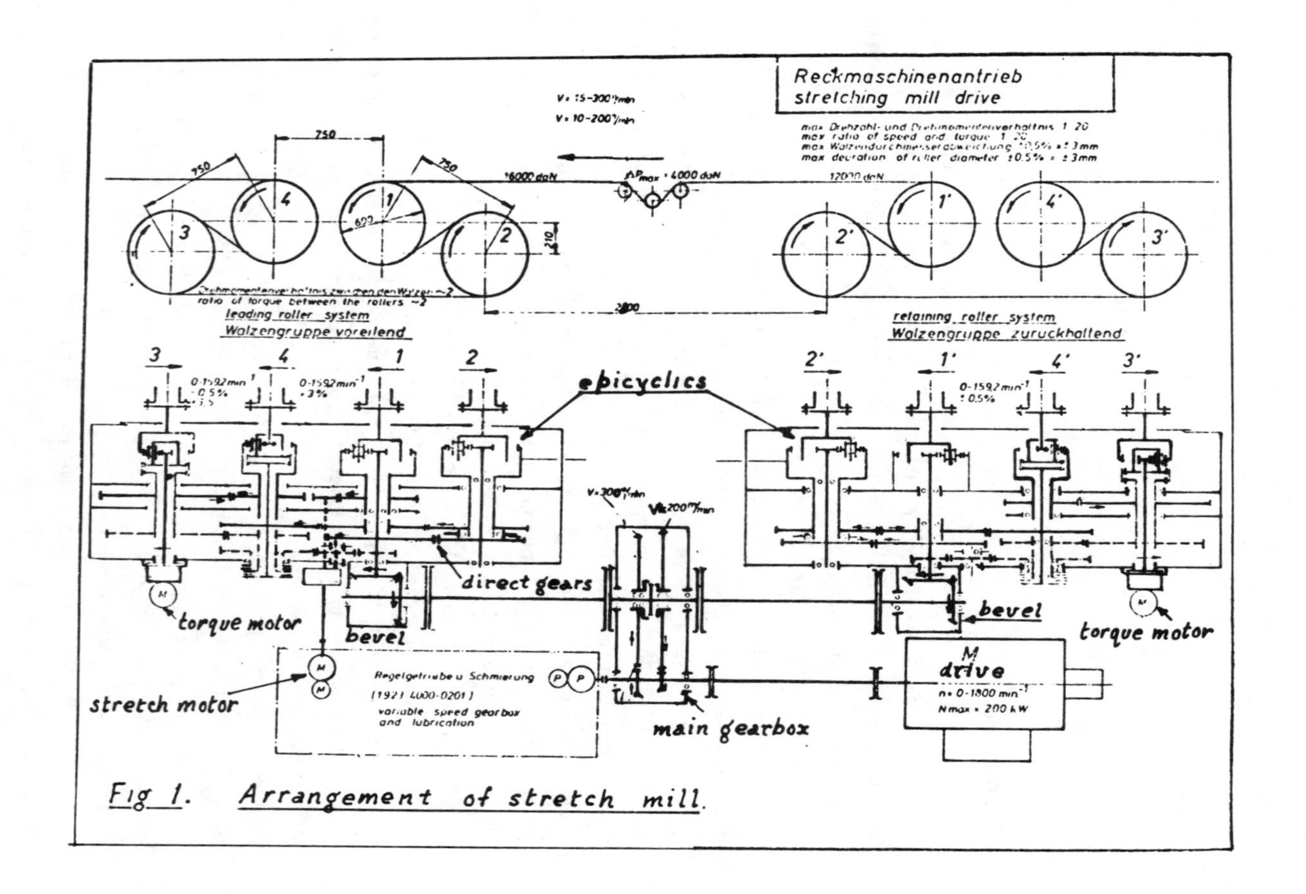

Fig 1. Arrangement of stretch mill.

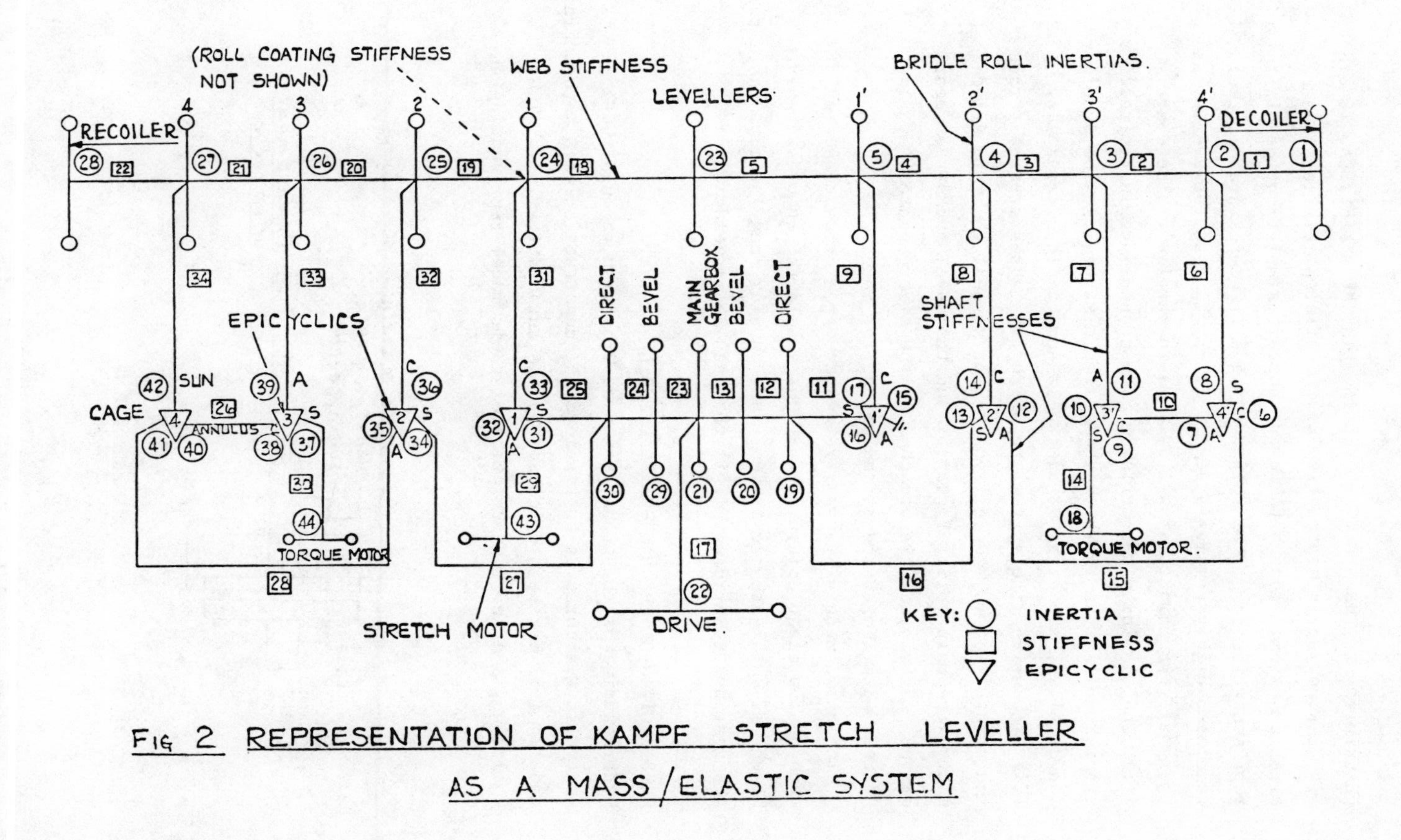

FIG 2 REPRESENTATION OF KAMPF STRETCH LEVELLER AS A MASS/ELASTIC SYSTEM

DESCRIPTION OF THE STRETCH LEVELLING LINE AS A MASS/ELASTIC SYSTEM

Fig. 2 shows an interpretation of the stretch leveller as a mass/elastic system. Though there are certain simplifications in this representation (e.g. the roll coating stiffnesses are omitted), it remains a highly complex system. It may be seen, however, that there are two main torsional systems in parallel. These are:

(i) The aluminium web acting as the connecting elasticites for a torsional system in which the multiple masses are the recoiler and decoiler, the eight bridle rolls and the leveller rolls.

(ii) The drive line system, in which the masses are the torque and stretch motors, the epicyclic gear masses, the main drive motor, the main splitter gearbox and the bevel and direct gear drives. This system is branched at the stretch motor, and the "direct" gear boxes and at the main drive motor.

These two main vibratory systems are then cross connected at each of the eight epicyclic gear boxes. In the epicyclic gear boxes, all elements are driven (sun, annulus, cage) except that in the gear box to bridle roll 1' the annulus is held.

CONSTRUCTION OF THE MATHEMATICAL MODEL

Understanding of the transmission of vibration through the eight epicyclic gearboxes is critical to the construction of a mathematical model. This has been discussed by the authors (ref. 1). The following is a short description of their model representing an epicyclic gearbox.

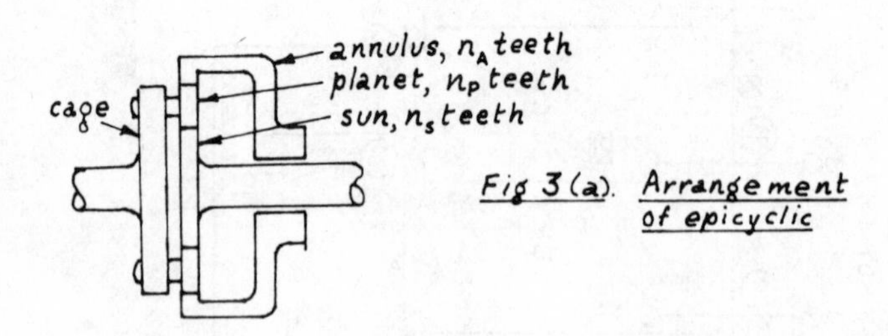

Fig 3(a). Arrangement of epicyclic

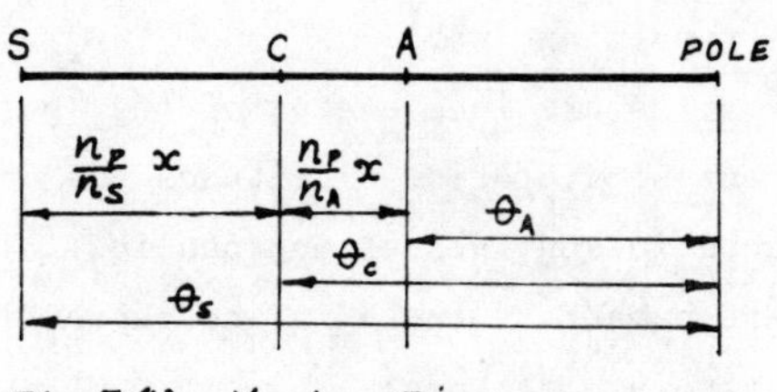

Fig 3 (b). Vector Diagram

Relative motion of the components may conveniently be represented by a vector diagram.

Since the gearbox systems is linear, θ may represent any time derivative function, such as displacement, velocity, acceleration. x is a scale factor.

Considering torques, two conditions must be met

Σ Torque = 0

Σ Power = 0 i.e. $\Sigma T\dot{\theta} = 0$

and if d'Alembert's principle is applied, the equations are valid for dynamic situations.

A useful analogy applies torques to the vector diagram representing velocity as shown in Fig. 4 below. This is generally referred to as the beam analogy.

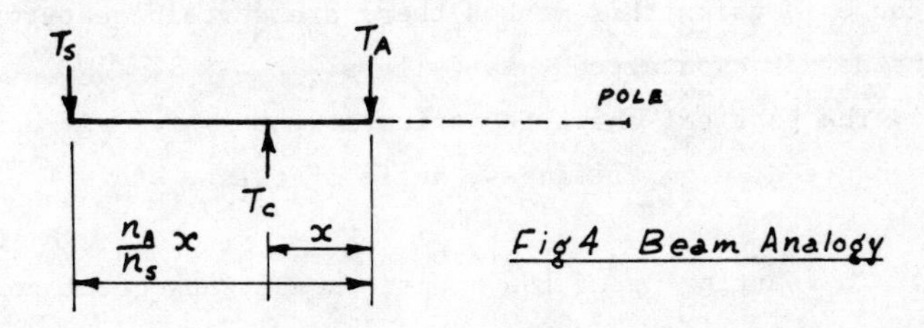

Fig 4 Beam Analogy

It can easily be shown that the torque and power equations are satisfied.

In analysis of vibration in an epicyclic gearbox, when cage, sun and annulus are all free to move,

(1) Two of the three vibration amplitudes must be known before the third can be evaluated.

(2) One of the three torque amplitudes must be known in order to evaluate the remaining two.

(3) Torque ratios are independent of the "pole position" i.e. they are independent of vibration amplitudes and depend only on the number of teeth in sun, planet and annulus.

If n is the number of teeth in the respective gears:

(a) for vibratory amplitude:

$$\theta_C = \theta_A + \frac{n_p}{n_a} * \chi$$

$$\theta_S = \theta_A + \left(\frac{n_p}{n_a} + \frac{n_s}{n_p}\right) \chi$$

where subscripts p, s and a refer to planet, sun and annulus respectively and χ is an unknown scale factor.

(b) for vibratory torque:

$$T_C = -\left(1 + \frac{n_s n_a}{n_p^2}\right) \times T_S$$

$$T_A = \frac{n_s n_a}{n_p^2} \times T_S$$

Several approaches to the solution of the problem can, of course, be considered. A development of the Holzer approach was adopted because in using this method there are certain features which were considered advantageous, as follows:

i) The physical characteristics of the system are easily recongnized, e.g. stiffness, angle of twist, etc., for particular members.

ii) The setting up of the equations and subsequent computer flow chart, is progressive through the mass/elastic system. Each step can be checked for accuracy, indeed, sub-sections of the system can be analysed independently, a very necessary facility in dealing with a mechanism of this complexity. Such sub-systems can be isolated as sub-routines in a computer programme.

(iii) The mode shapes of the vibration are very quickly drawn once the eigenvalues have been determined.

(iv) After determination of eigenvalues, non-resonant or near resonant forcing torques can be applied at any chosen position or positions in the mechanism.

(v) The method provides information throughout the frequency spectrum; for example, it was possible to identify broad bands of frequency at which the system was close to resonance, that is, bands within which a very light forcing might be expected to produce relatively large amplitudes.

The basic Holzer table is not, of course, capable of dealing with a system of the complexity of this stretch line. Iterative computation techniques were developed to achieve compatibility at the common points of the branches and cross connections, and by this means the calculation was carried progressively through the system. It was found necessary to split the system near the centre plane and to work to this common plane from each end, subsequently achieving compatibility at this junction by re-iteration of the complete calculation from one of the ends, while varying the arbitrarily chosen input conditions.

A much fuller treatment of the developed approach follows later in the discussion. At the moment, we are considering the broad philosophy.

Analysis at the epicyclic gearboxes was by the methods already described, requiring as input one known torque and two known amplitudes.

The concept of dynamic stiffness is used in achieving total compatibility at the common plane in the web line and the calculation is then progressed to achieve compatibility of displacement angle at the common plane in the drive line, while maintaining the full matching of the web line. When this has been achieved, a situation has been reached in which only the angular displacement input at the decoiler end remains arbitrary and the only remaining incompatibility is in torque at the common plane in the drive line. This whole calculation may now be iterated against frequency. Any frequency resulting in a torque difference across the common plane at the drive line (a residual torque) is a non-eigenvalue, but any frequency which can be identified as giving matched torques at the common plane in the drive line becomes completely compatible, that is, the system is capable of free vibration and we have identified an eigenvalue or natural frequency.

This method was found to be satisfactory over most of the range of frequencies between about 15 Hz and 120 Hz. Certain narrow frequency bands were, however, identified within the range, in which the iteration procedure became unstable and compatibility at the junction could not be achieved. Thorough examination of this condition showed that instability at certain frequencies was inherent in the nature of the parameters of the junctions and could not be avoided. At least one of these instabilities was close to a region of rippling frequency and it was therefore considered essential to close these gaps in the residual torque/frequency diagram. To this end, a second programme was developed along similar lines to the first, but with the junction line moved from the central position to a position near the recoiler. This programme was also subject to instability at the common line, but at different frequencies, so that between the two programmes, it was possible to construct a complete diagram of residual torque. This second programme was developed independently and therefore served also as a check on the correct compilation of the first programme, in that within the broad bands of frequency within which both programmes were stable, the mode shapes and eigenvalues were identical. The complexities of compatibility are dealt with in more detail under a later heading.

In very broad outline, the first stage of the computation proceeded along the indicated lines in Appendix I, to evaluate dymanic stiffness of the web line on each side of the common plane.

APPLICATION OF THE CONCEPT OF DYNAMIC STIFFNESS TO THE STRETCH LEVELLER

Notes on the principle of dynamic stiffness are included in Appendix II.

Preliminary work on the stretch levelling line had shown that it would not be feasible to carry a Holzer type calculation all the way through the twin lines, to give a final residual torque at the recoiler end. After considerable investigation, it was concluded that the most efficient approach would in fact be to split the system near the centre plane, on the recoiler side of the levellers in the web line and of the main drive motor and main gearbox in the drive line, then to match these systems at the common plane by the use of the dynamic stiffness principle. In the end, it was found necessary to write two programmes, one having a common plane as described, while the other represented a split at roll 3 (inertia 26) in the web line and at the annulus of gearbox 2 (inertia 35) in the drive line. This dealt with the problem of instability of the calculation when using those frequencies at which a node falls at the common plane. The stretch line is not, of course, symmetrical about the central plane, mainly because of the stretch motor at the recoiler end. However, there are many similarities between the recoiler and decoiler ends, and this considerably eased the programming load.

The following discussion is written as for the first of these programmes, but the argument is identical for both programmes, with the substitution of appropriate inertias and stiffnesses.

As outlined in the previous discussion and flow chart, the calculation is initiated at the decoiler end by an arbitrary angle input at (1), (18) and (12), say θ_1, θ_{18}, θ_{12}, together with an arbitrary torque input at (12), say T_{12}. The calculation then proceeds through the system, using Holzer methods and the newly introduced analysis of vibration in epicyclic gearboxes, to arrive finally at the common plane on the recoiler side of inertia (23) and inertia (21). In the course of this calculation, iteration to secure compatibility has adjusted θ_{18} and θ_{12}, so that the remaining arbitrary values of input are θ_1 and T_{12}. The outputs of this calculation are torque and angle at (23), say T_{23}, θ_{23} and torque and angle at (21), say T_{21}, θ_{21}. Dynamic stiffness on the web line form, the decoiler end may now be expressed as T_{23}/θ_{23}.

A similar calculation from the recoiler end yields torques and angles at the same positions with remaining arbitrary inputs θ_{28} and T_{35}.

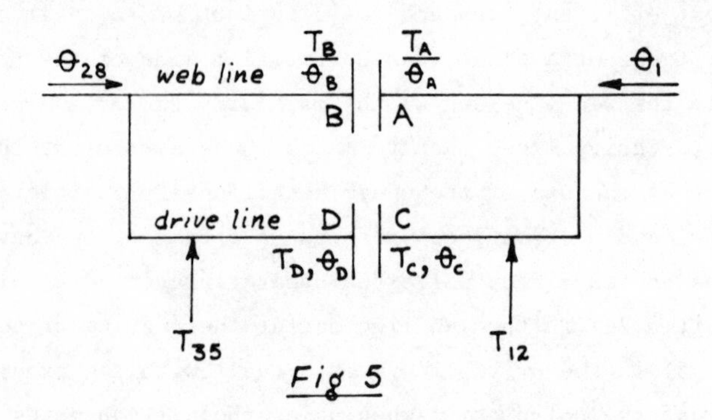

Fig 5

For clarity, refer now to Fig. 5 in which A and B are the common plane position on the web line, as approached from different sides and similarly C and D are the common plane positions on the drive line, as approached from different sides.

Thus far we have calculated torque and angle at each of A, B, C and D, none of these being at this stage compatible. The remaining inputs are θ_{28}, T_{35}, θ_1 and T_{12}.

From this stage we proceed as follows.

1. Calculate dynamic stiffnesses (T/θ) at A and B in the web line.
2. Iterate the calculation of DS_B (T_B/θ_B) while varying T_{35}, to find the value of T_{35} at which $DS_B + DS_A = 0$, i.e. match the dynamic stiffnesses. In practice a straightforward arithmetically progressing iteration of this calculation is impossibly time consuming and it was found necessary to develop a technique for rapid convergence on the zero error value. This in itself was a considerable task and it is therefore described separately under a later heading.
3. Proportion torques and angles on the recoiler side (i.e. the B, D side) to give a separate match of torque and angle in the web line, i.e.

$$T_B = -T_A$$

$$\theta_B = \theta_A$$

The web line is now fully matched. It remains to match the drive line.

4. In the above condition, measure the displacement error at the C-D interface, i.e. $\theta_D - \theta_C$. The entire calculation to this stage should now be iterated while varying the sole remaining arbitrary input T_{12} (other than end displacement θ_1) thus giving error $(\theta_D - \theta_C)$ against T_{12}. The desired value of T_{12} is that which gives $\theta_D = \theta_C$, i.e. zero error. The matching at the A-B interface is re-established during each pass of this calculation. While programming to this end was in progress, it was noted that the variation of error $(\theta_D - \theta_C)$ against T_{12} was totally linear. While no completely satisfactory explanation has yet been found for this linearity, it was nevertheless gratefully received and used to predict the zero error value of T_{12} from only two passes of the calculation. The linearity is, of course, checked by the insertion of the predicted value of T_{12}, and calculation of error $(\theta_D - \theta_C)$, in other words, the final calculation is identical whether arrived at laboriously by iteration, or by using the linearity short cut.

At this stage in the calculation, the only remaining arbitrary input is the customary Holzer initial displacement input θ_1.

5. We have now achieved a full match on the web line and a match of angle at the common plane in the drive line. The remaining incompatibility lies in the torques on the drive line at the common plane.

If now the entire calculation is iterated with arithmetically progressing vibration frequency, we can identify the torque difference at the drive line common plane as a "residual torque". At certain frequencies this torque difference is zero, implying that the system is fully matched on both lines. In such cases, the system is capable of free vibration as a whole system, in other words, these frequencies identify the eigenvalues of the system.

The corresponding flow chart for these processes is shown in Fig. 6.

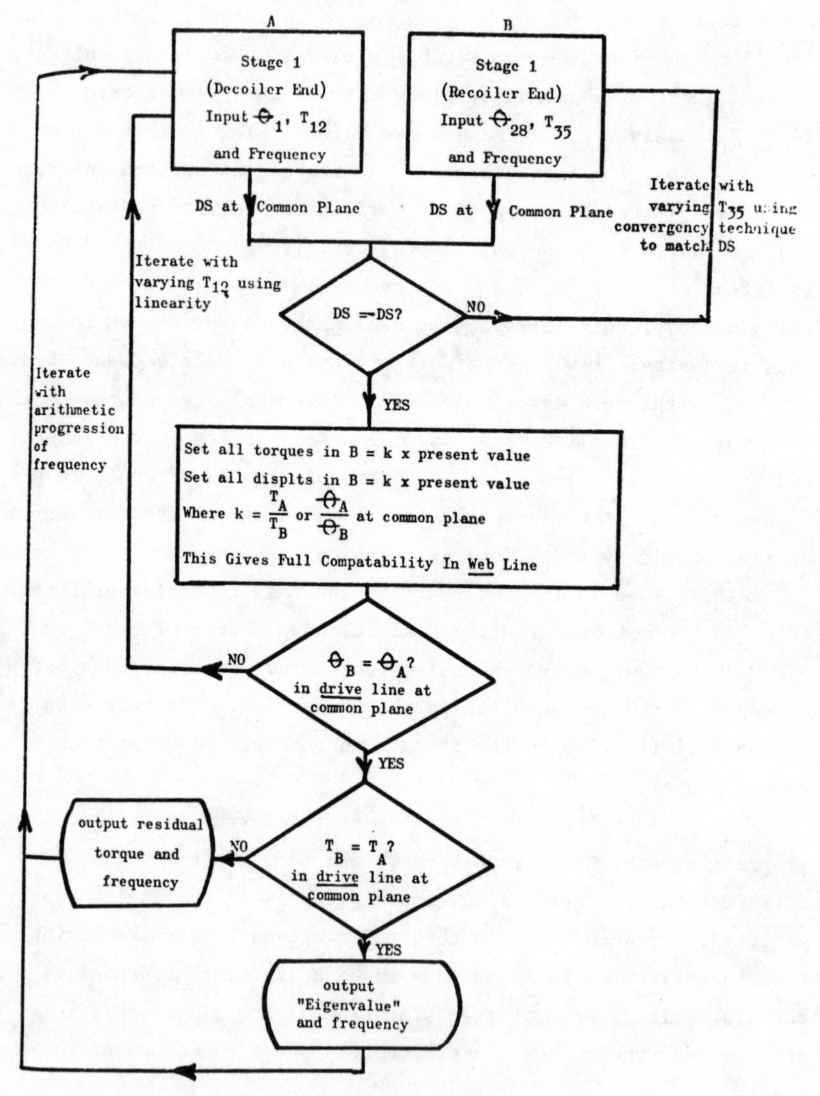

FIG 6 FLOW CHART OF MATHEMATICAL MODEL OF STRETCH LINE
(see also detailed flow chart of Stage 1) (Appx I)

CONVERGENCY TO ZERO ERROR IN DYNAMIC STIFFNESS

Those familiar with the matching of torsional systems by the use of the dynamic stiffness concept will recognise the curves of Fig. 7.

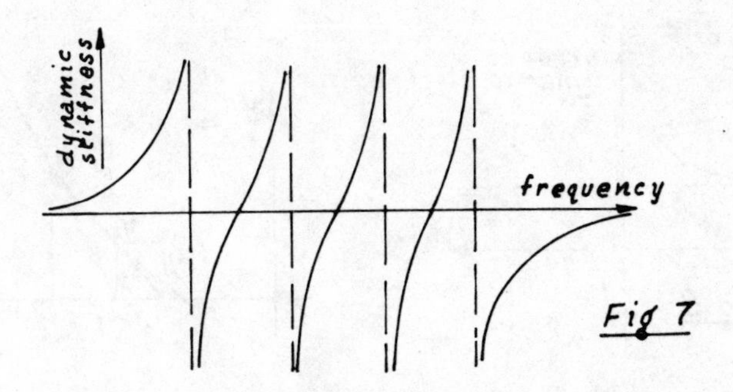

Fig 7

These are curves of dynamic stiffness for a notional torsional system at the interface with a second torsional system, plotted to a base of frequency. They are commonly known as tuning curves.

The crossings of the x-axis identify the eigenvalues of the individual system, while intersections with similar curves for the second system identify eigenvalues of the total coupled system. As frequency changes, the positions of the nodes in the separate systems change and at a change of order, an additional node appears following a reduction of amplitude at the coupled face to zero, in one or other of the coupled systems. At this change, dynamic stiffness on that side of the interface (T/θ) becomes infinite. This leads to a vertical asymptote in the tuning curves at a change of order. As frequency tends to zero or to infinity, dynamic stiffness tends to zero, giving a horizontal asymptote as shown.

The application of the dynamic stiffness concept to the matching of the web line of the stretch leveller has similarities to this tuning operation, but is by no means identical. Here dynamic stiffnesses are compared at the web line interface, but with an x-variable of torque input at (35), while frequency is held constant. This places severe restraints on the ability of the system to change order. Careful examination suggests, and experiment confirms, that in these circumstances, the sub-system can change order only once, following the arrival of a node at the interface. The curve of error in dynamic stiffness to a base of torque input at (35) thus has the form of fig 8,

8.,

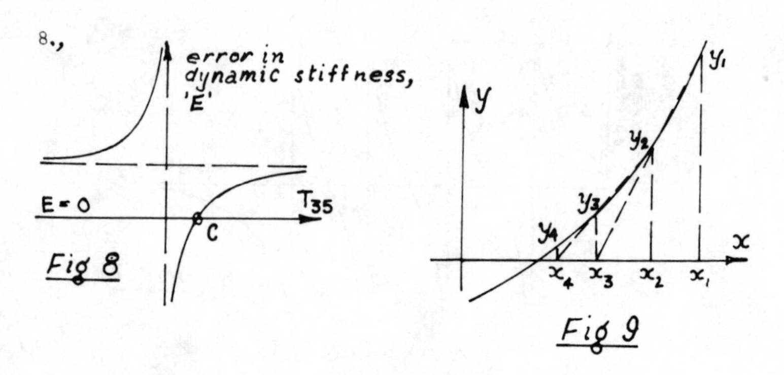

Fig 8

Fig 9

which will be recognised as having close similarities to the first and last branches of the tuning curves previously described. Depending on frequency, this branched curve may occupy either the 2nd and 4th quadrants as illustrated, or the 1st and 3rd quadrants. Since the input is a torque on one side of the interface only, the error does not tend to zero at high and low values of input, but rather to a steady value representing the difference of the fixed dynamic stiffness on one side of the interface, and the steady dynamic stiffness on the other side induced by massive positive or negative torque inputs at (35). In other words, the dynamic stiffness when the sub-system vibration is dominated by the input torque. The curve thus has a vertical and a horizontal asymptote and the problem is that of identifying, without massive arithmetically progressing iteration, the point C in Fig. 8 that is to say, the value of torque input at (35) to produce zero error in dynamic stiffness accross the interface.

The method used in iterating the programme to home on zero values is an extensive development of a simple and well known technique of gradient extra-polation. In Fig. 9 we need to know two values of y, say y_1 and y_2, at x_1 and x_2. x_1 and x_2 may be chosen arbitrarily (or by educated guess) then two iterations of the calculation yield y_1 and y_2. If the gradient of the line joining points 1 and 2 is now extrapolated to the crossing with the x-axis we can iterate with x_3 to find y_3. A further pass using (x_2, y_2), (x_3, y_3) as initial values yields (x_4, y_4) and so on. This procedure converges very rapidly on a true $y = 0$ value, provided y is a single valued smoothly varying function of x.

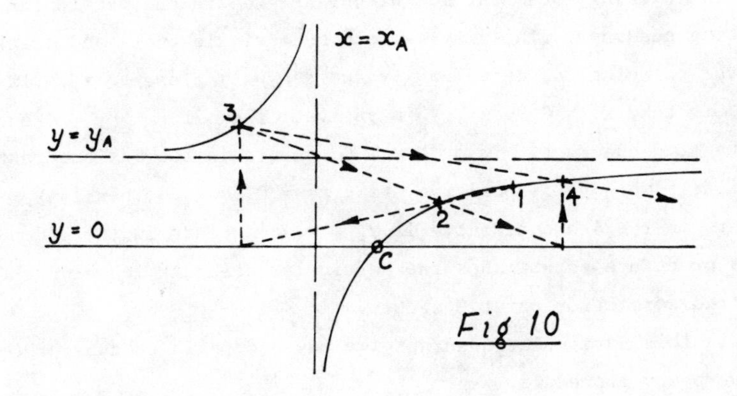

Fig 10

In the stretch line mathematical model, iteration is carried out to match dynamic stiffness in the web line at the selected common plane. The curve with which we have to deal is thus a curve of error in dynamic stiffness on one side of the common line, to a base of torque input T_{35}. As discussed above, the error curve in the stretch line was found to be of the type shown in Fig. 10. The curve shows a vertical and a horizontal asymptote at $x = x_A$ and $y = y_A$ respectively, though these asymptotic values are not known at this stage. Following the gradient extrapolation method through points 1, 2, 3 and 4 in the figure illustrates the fact that the basic method of homing to the $y = 0$ value becomes unstable when the extrapolated gradient crosses an asymptote and joins the other branch of the curve. The required point C occurs on whichever branch of the curve crosses the x-axis, but this branch may occur in any one of the four quadrants, since the $y = y_A$ asymptote may be either above or below the $y = 0$ line.

Before application of the convergency technique, the following procedures are therefore required.

1. The x-value of the vertical asymptote and the y-value of the horizontal asymptote must be established. If y_A is positive, the crossing point C will occur in the third or fourth quadrant. If y_A is negative, C will occur in the first or second quadrant. The method used to determine these asymptotes is fairly complex and will be described in a section following the present discussion.

2. The working area must be further defined to the particular working quadrant. This may be done by a single guess and trial pass of the calculation, using an x-value (say x_g) close to x_A. In Fig. 11 take $x_g = 0.99\ x_A$, y_A negative. Then if $y_g > y_A$, C occurs in the 2nd quadrant. If $y_g < y_A$, C' occurs in the 1st quadrant. Similarly when y_A is positive, take $x_g = 0.99\ x_A$. Then if $y_G > y_A$, C occurs in the 4th quadrant. If $y_g < y_A$ C' occurs in the 3rd quadrant. This procedure establishes the particular quadrant in which the desired zero error point C occurs.

3. In this particular quadrant, we may now begin to apply the convergency procedure.

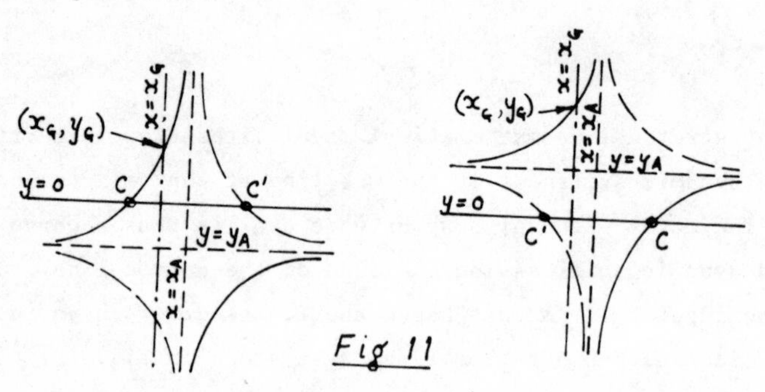

Fig 11

Take two values of x close to the asymptote $x = x_A$ (one may already have been calculated in some cases) and iterate the calculation to initiate the convergency procedure. This will work well in most cases. However, at large values of y_A and/or at large x, the term "close to the asymptote $x = x_A$" requires examination. In such circumstances, it is still possible to select initial values such that C lies between these values and the vertical asymptote. If this happens the extrapolated gradient will still cross the aymptote, leading to instability in the convergency procedure. This possibility is easily, if rather inelegantly, cured by including a sub-routine in the programme to prevent any attempt to cross the asymptote, that is to say, if at this stage of the convergency procedure the extra polated gradient crosses the asymptote, the convergency is discontinued, then re-initiated with fresh initial values of x closer to the asymptote than those first used. This procedure may, if necessary, be repeated until initial x-values are between C and the asymptote, leading to a stable convergency to y = 0.

With these considerable modifications, the convergency procedure works well for all conditions, with one exception. As frequency is increased, the nodes change position in such a way as to prepare for the change to the next node of vibration (see simple example in Fig. 12).

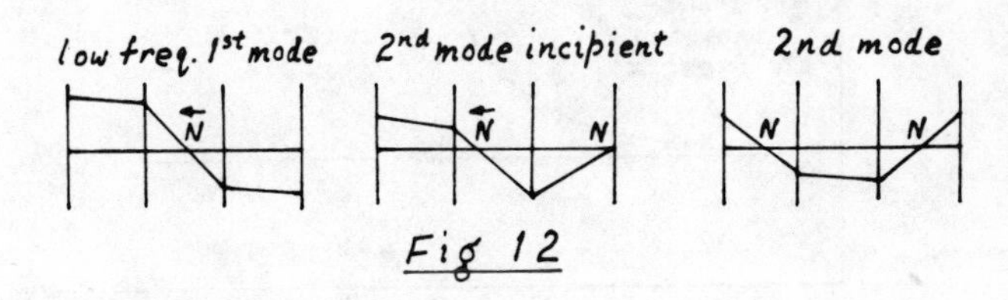

Fig 12

In these circumstances, frequencies inevitably occur at which the node falls exactly on the inertia chosen as the common plane for the matching of dynamic stiffness. Since dynamic stiffness is external torque/amplitude, this means that dynamic stiffness becomes infinite on the common plane and the horizontal asymptote of the above discussion is removed to plus or minus infinity. Naturally, no convergence procedure will work in these circumstances and since the phenomenon is inherent in the mechanics rather than in the mathematics of the system, there is really nothing that can be done about it. It was to cater for such gaps in the calculation of dynamic stiffness error that two programmes were set up, with common planes in the centre plane of the system, and near one end, respectively. While the last unavoidable instability described above occurs in both programmes, it occurs at different frequencies and it is thus possible to cover the whole range of frequency. The existence of unspecified eigenvalues within the scanned frequency range is detected by constructing the normal elastic curves for each eigenvalue and determining the number of nodes. Successive eigenvalues should display one additional node on the normal elastic curve.

DETERMINATION OF THE ASYMPTOTES

The convergency procedures used in the mathematical model has general application to the determination of zero values in branched curves. For this reason, and because the general approach tended to be simpler in this case, the discussion has been presented in a general cartesian x-y field. The asymptotes, however, are particular to the hardware under examination, so we must return to the stretch leveller to discuss the methods used in their determination.

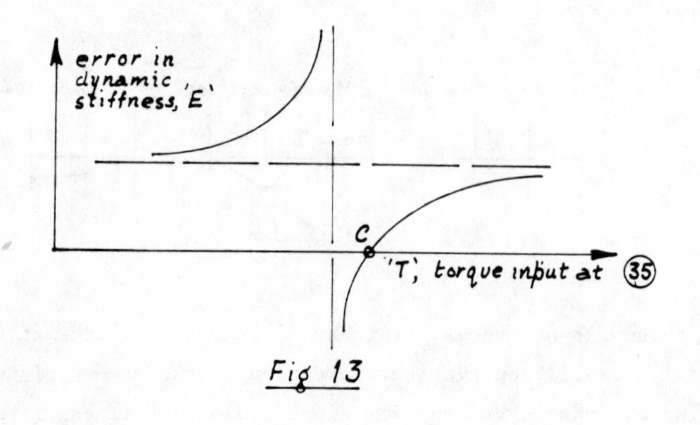

Fig 13

In Fig. 13, the branched curve is presented in its real co-ordinates of error in dynamic stiffness at the common line as ordinate, versus torque input at (35) as abscissa. It is clear that the horizontal asymptote is fairly easy to determine, being simply the E value at large positive or negative T. One pass of the calculation with $T = \pm 10^{10}$ is sufficient to determine this value. Solution of the vertical asymptote is, however, an altogether more difficult problem. E compares the dynamic stiffness on one side of the common line, as T varies, with the fixed dynamic stiffness on the other side of the common plane at a particular frequency. This being so, it is clear that the vertical asymptote occurs when the varying dynamic stiffness goes to infinity. Dynamic stiffness being the ratio torque/displacement at the common line, this implies that displacement has gone to zero, in other words, the vertical asymptote represents that value of input torque at (35) which brings the web line node to the common plane. For the programme with common plane at (26) the required condition is zero displacement at (26). In broad terms, the approach taken to this problem is to calculate separately the contributions to θ_{26} of the web line and of the drive line, then so to amend the torque at (35) that the drive line contribution is equal and opposite to the web line contribution, thus bringing the net displacement to zero. This may now be examined in more detail.

THE WEB LINE CONTRIBUTION

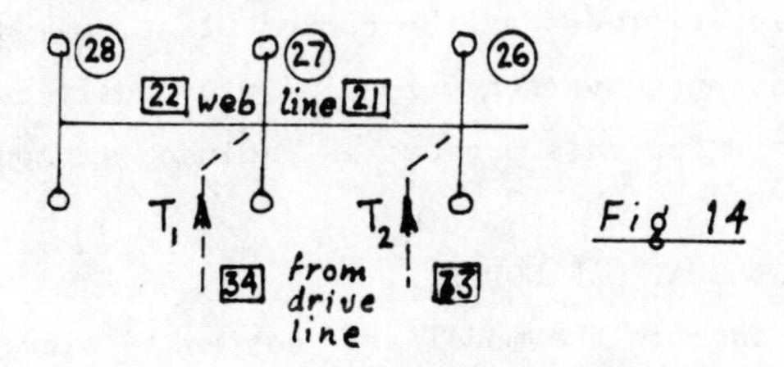

Fig 14

To evaluate the web line contribution, the input torques on the drive line are made zero. Taking now a Holzer treatment along the web line, we input our arbitrarily chosen initial displacement at (28), say θ_{28}. Since the torque contributions along shafts [34] and [33] are now zero, the calculation becomes a straightforward single line three mass Holzer, yielding a displacement, say θ_W at (26). This is the contribution at (26) of the web line alone.

In a second pass of the calculation, we input a large torque at (35) say T_{35} and reduce the displacement input at (28) on the web line to zero. The input torque is chosen large, to dominate the contribution from the drive line, the contribution from (44) being relatively small. Consider first the effect of this on the web line, that is, consider the full line diagram only. A simple Holzer calculation shows that if θ_{28} is zero, θ_{27} and θ_{26} are also zero, in other words, there is no contribution to θ_{26} from the web line.

Taking now the whole system and applying Holzer methods, an input of $\theta_{28} = 0$ again gives $\theta_{27} = 0$. The onward transmitted torque, however, is not now zero, because of the torque contribution along line [34], arising from torque inputs at (44) and (35). This imparts a "twist" to line [21], leading to a displacement, say θ_D, at (26). Clearly θ_D is the contribution of the drive line to θ_{26} and $\theta_{26} = \theta_W + \theta_D$ when the inputs are θ_{28} and T_{35}.

T_{35} may now be proportioned to make $\theta_D = -\theta_W$, thus making $\theta_{26} = 0$. With these inputs, dynamic stiffness becomes infinity and this defines the T_{35} value for the vertical asymptote.

The treatment for the other programme, with the common plane at (23), is similar.

The asymptotic values of T, the torque input at (35), and of E, error in dynamic stiffness at the common plane, may now be inserted as x_A and y_A in the convergency routine previously described.

The flow chart for this process is included in Appendix I.

RESULTS FROM MATHEMATICAL MODEL

The parameters used in the model are notional since it was not possible to obtain accurate inertia and stiffness values. This applies particularly to the stiffness of the hydraulic fluid lines of the torque motors. The results obtained are therefore qualitative rather than quantitative, but the conclusions are obviously valid and were accurate stiffnesses and inertias to be inserted, we have confidence that the quantitative output would then be reliable.

Fig. 15 shows the residual torque diagram. The eigenvalues are located where the curve crosses the zero residual torque axis.

Fig. 16 shows a typical normal elastic curve, this example being for the 4th normal mode of vibration.

It is interesting to note in Fig. 15 that comparatively wide bands of frequency exist (e.g. 70 to 90 Hz) in which the residual torque is comparatively small. Under these circumstances, near resonant vibration obviously may occur with small forcing, particularly if system damping is low. Such zones occupy a large proportion of the speed range. Such zones are present because, when the forcing frequency is altered, the velocity ratios of the 8 epicyclic boxes alter (see Analysis of Epicyclic Gears). We thus have a system with variable parameters, which unexpectedly produces this tendency to dwell near resonance. This will, of course, be characteristic of all machines of this type and is not affected in principle by the accuracy of the input parameters.

Examination of the complete set of normal elastic curves shows that the levellers in most circumstances seem to be comparatively close to a node.

REFERENCES

1. Vibration Analysis of Epicyclic Gears. R. Davidson, J. Brown. Conference on Teaching of Vibration and Noise in Higher Education, University of Sheffield, 1979.

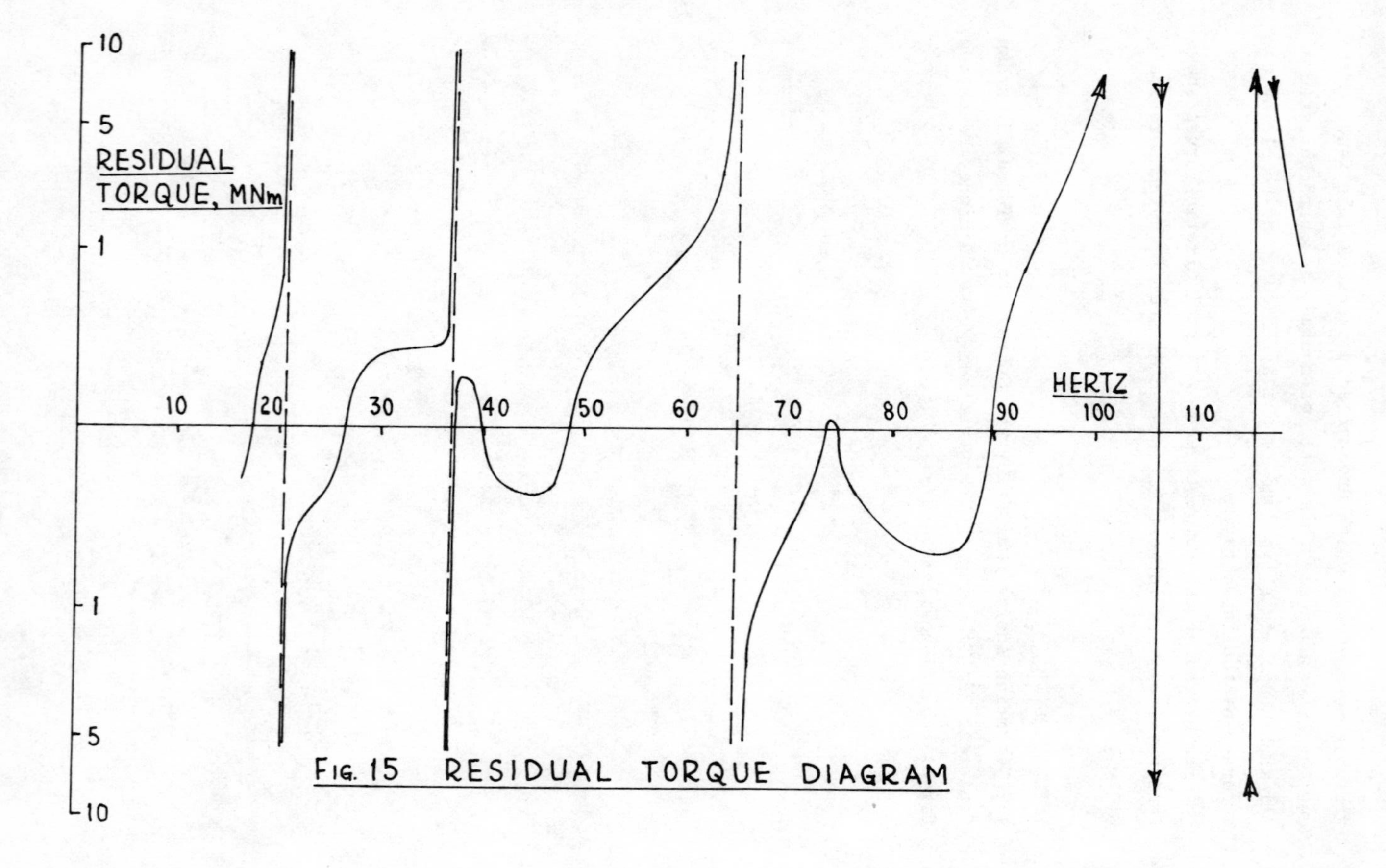

FIG. 15 RESIDUAL TORQUE DIAGRAM

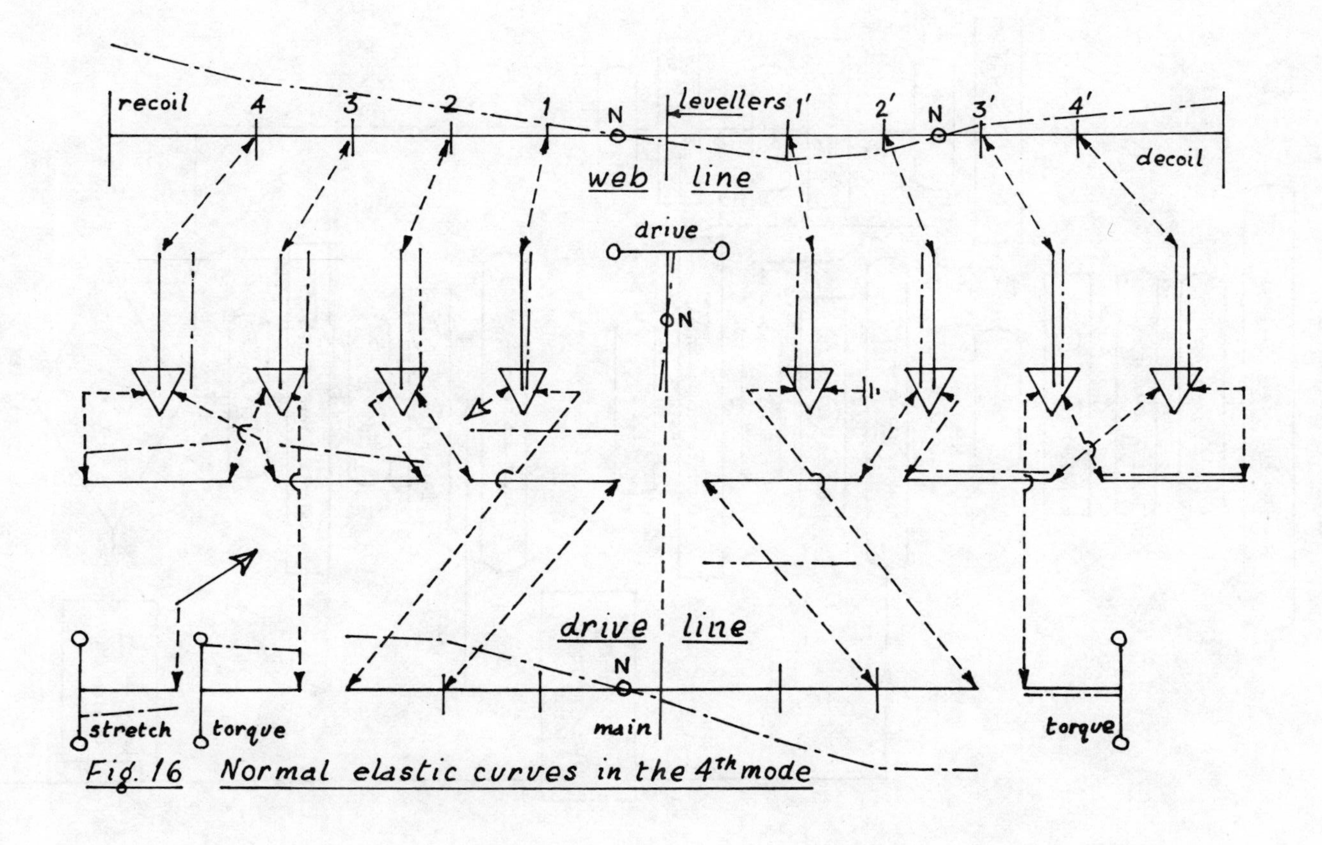

Fig. 16 Normal elastic curves in the 4th mode

APPENDIX I - FLOW CHARTS

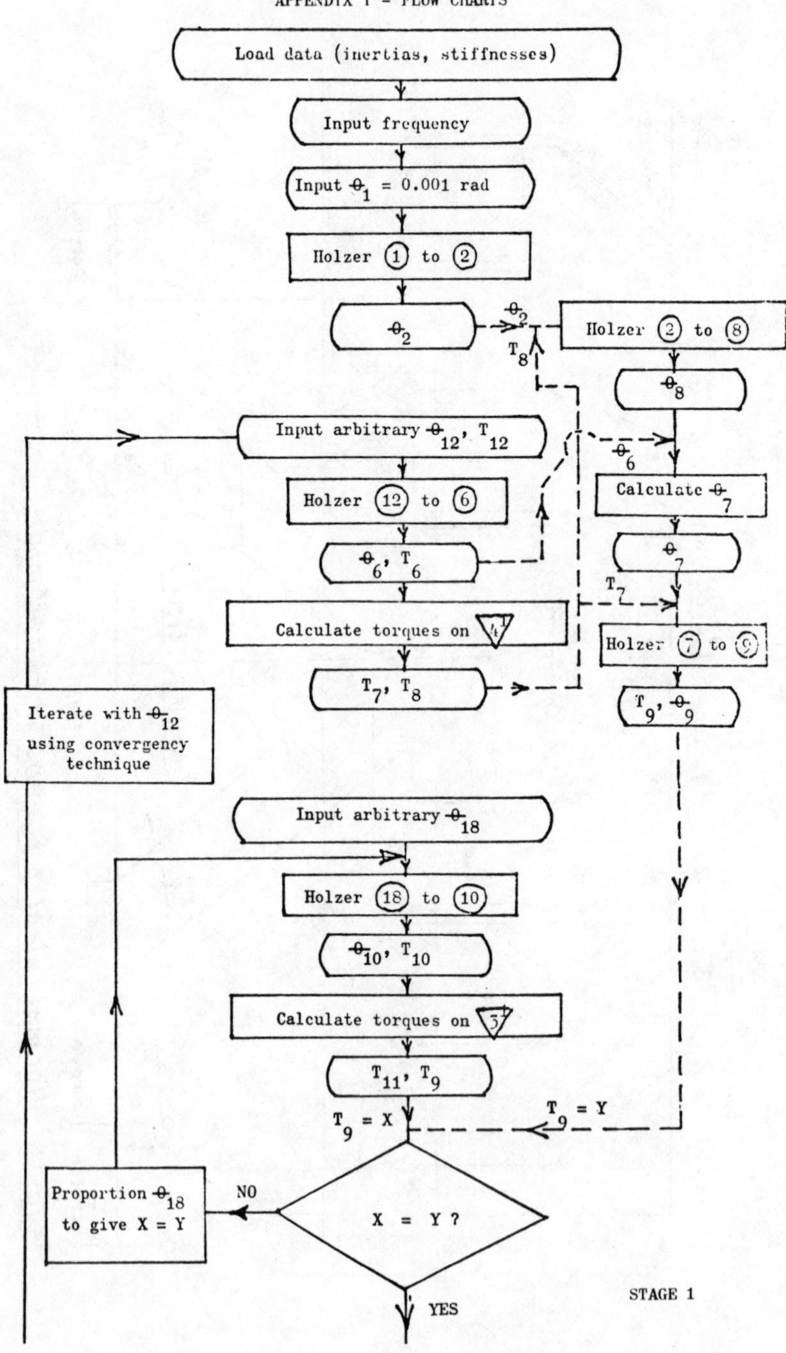

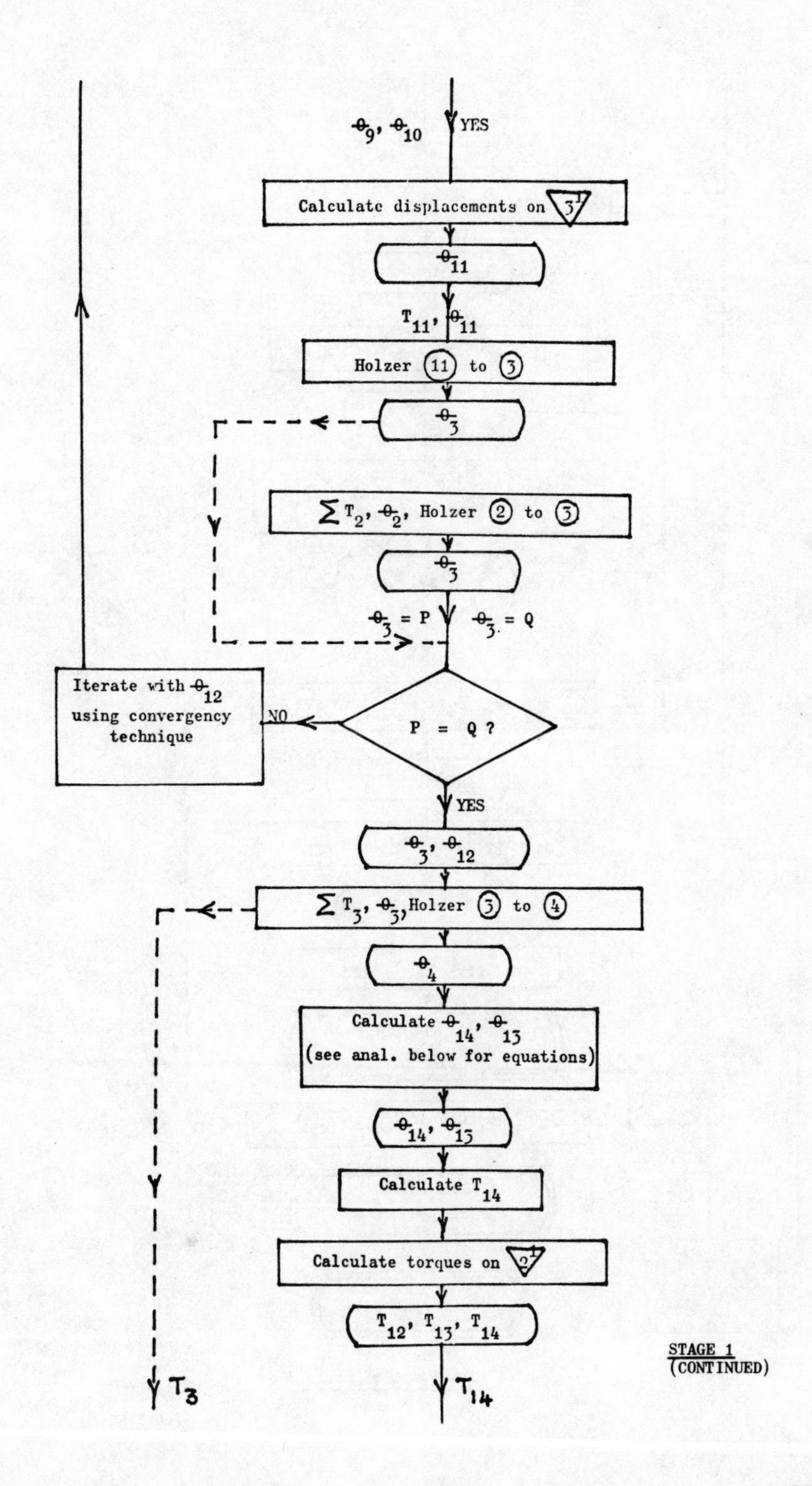

YES
θ_9, θ_{10}
Calculate displacements on 3[1]
θ_{11}
T_{11}, θ_{11}
Holzer (11) to (3)
θ_3
$\sum T_2$, θ_2, Holzer (2) to (3)
θ_3
$\theta_3 = P$
$\theta_3 = Q$
Iterate with θ_{12} using convergency technique
NO
P = Q ?
YES
θ_3, θ_{12}
$\sum T_3$, θ_3, Holzer (3) to (4)
θ_4
Calculate θ_{14}, θ_{13} (see anal. below for equations)
θ_{14}, θ_{13}
Calculate T_{14}
Calculate torques on 2[1]
T_{12}, T_{13}, T_{14}
T_3
T_{14}
STAGE 1 (CONTINUED)

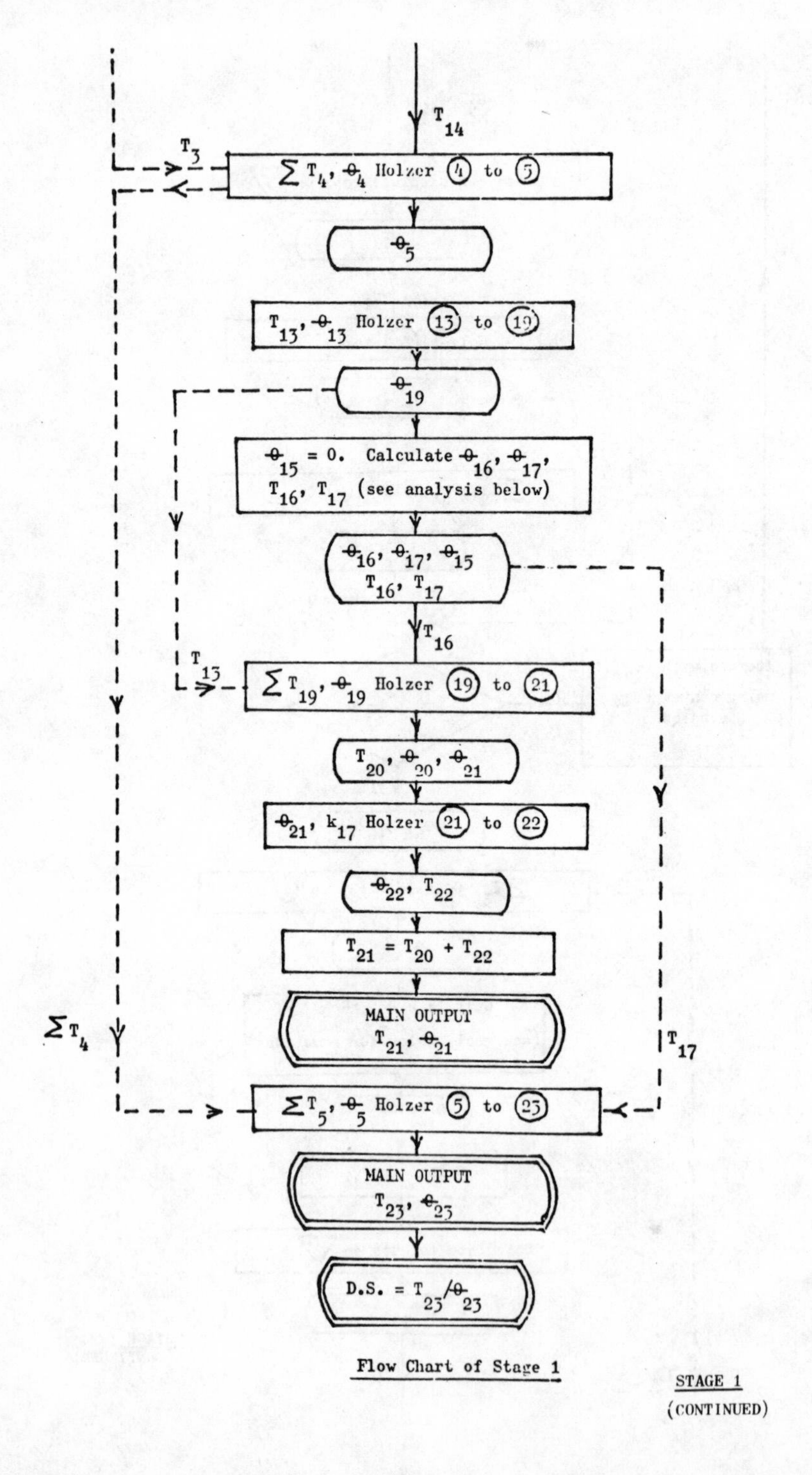

Flow Chart of Stage 1

STAGE 1
(CONTINUED)

ANALYSIS USED IN STAGE 1 OF THE MATHEMATICAL MODEL

(a) Standard Holzer Method

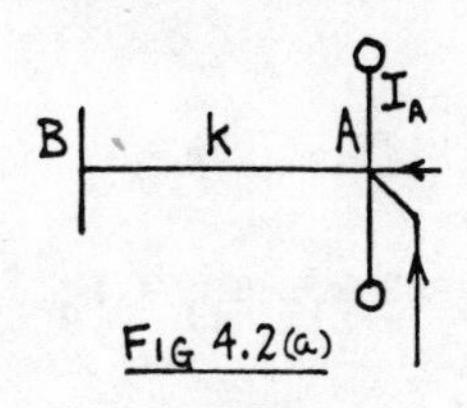

FIG 4.2(a)

Total torque at A = $I_A\, n^2 \theta_A$ + torques from previous stage

$$\text{Twist in shaft} = \frac{T_A}{k}$$

$$\theta_B = \theta_A - \text{Twist}$$

(b) Gearbox $\triangledown 2^1$

FIG 4.2(b)

θ_4 is known from preceding work,

$$\text{then} \quad \frac{T_{14} + I_{14}\, n^2 \theta_{14}}{k_8} = \text{Twist}$$

$$= \theta_{14} - \theta_4$$

$$k_8\theta_{14} - k_8\theta_4 = T_{14} + I_{14} n^2 \theta_{14}$$

$$\theta_{14} = \frac{T_{14} + k_8\theta_4}{k_8 - I_{14} n^2}$$

θ_{12} is known from preceding work,

Knowing θ_{12}, θ_{14}, calculate θ_{13} in the normal way, as per "Analysis of epicyclic gears." This completes analysis of gearbox $\triangledown 2^1$.

c) Gearbox $\triangledown 1^1$

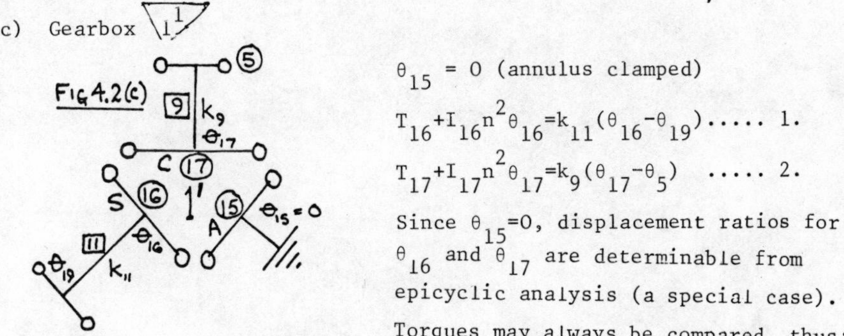

$\theta_{15} = 0$ (annulus clamped)

$T_{16} + I_{16} n^2 \theta_{16} = k_{11}(\theta_{16} - \theta_{19})$ ..... 1.

$T_{17} + I_{17} n^2 \theta_{17} = k_9(\theta_{17} - \theta_5)$ ..... 2.

Since $\theta_{15} = 0$, displacement ratios for θ_{16} and θ_{17} are determinable from epicyclic analysis (a special case). Torques may always be compared, thus:

$T_{17} = -3 \times T_{16}$

$\theta_{17} = \theta_{16}/3$

POLE

Hence, in (2), $-3\,T_{16} + (I_{17}\frac{^2\theta_{16}}{3}) = k_9(\frac{\theta_{16}}{3} - \theta_5)$ .... 3.

and, working simultaneous equations (1) and (3)

$$3\,T_{16} + 3\,I_{16}\,n^2\theta_{16} = 3\,k_{11}\,(\theta_{16} - \theta_{19}) \quad \ldots\ldots\ldots\ldots\ldots\ldots\ldots 1.$$

Sum (1) and (3) to give

$$n^2\theta_{16}\,(3\,I_{16} + \frac{I_{17}}{3}) = \theta_{16}\,(\frac{k_9}{3} + 3\,k_{11}) - 3\,k_{11}\theta_{19} - k_9\theta_5$$

Let $C = n^2\,(3\,I_{16} + \frac{I_{17}}{3})$, $D = \frac{k_9}{3} + 3\,k_{11}$, $E = 3\,k_{11}\,\theta_{19} + k_9\,\theta_5$

Then $\theta_{16}\,(C - D) = -E$

$$\theta_{16} = -\frac{E}{C - D}$$

$$\theta_{17} = \frac{\theta_{16}}{3}$$

$$\theta_{15} = 0$$

Also $T_{16} = (3\,k_{11}\,(\theta_{16} - \theta_{19}) - 3\,I_{16}\,n^2\theta_{16})/3$

$$T_{17} = -\,3\,T_{16}$$

This completes analysis of gearbox $\triangledown 1^1$.

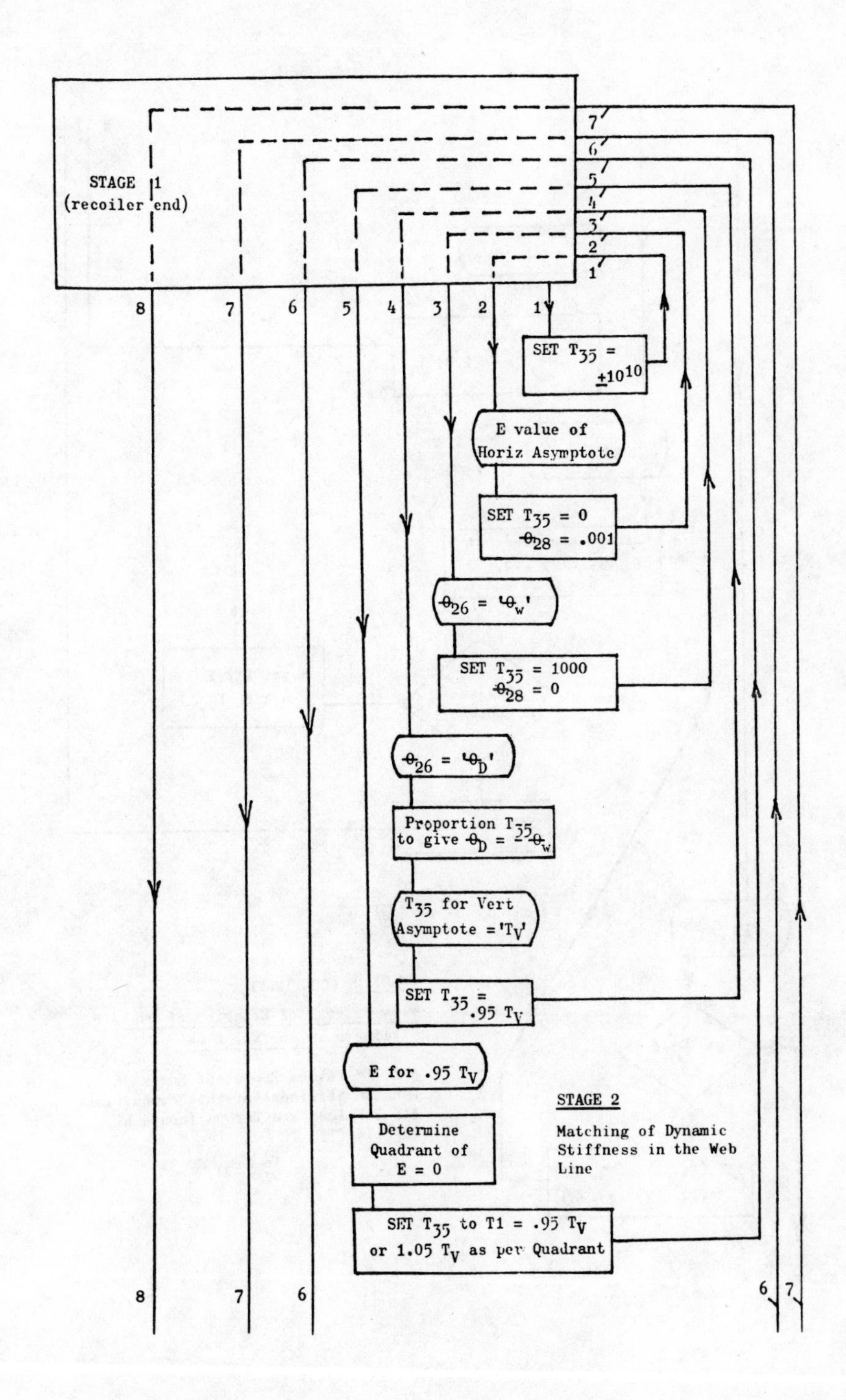
STAGE 1
(recoiler end)
7
6
5
4
3
2
1
8
7
6
5
4
3
2
1
SET T_{35} = $\pm 10^{10}$
E value of Horiz Asymptote
SET T_{35} = 0 θ_{28} = .001
θ_{26} = 'θ_w'
SET T_{35} = 1000 θ_{28} = 0
θ_{26} = 'θ_D'
Proportion T_{35} to give θ_D = 2 θ_w
T_{35} for Vert Asymptote = 'T_V'
SET T_{35} = .95 T_V
E for .95 T_V
Determine Quadrant of E = 0
SET T_{35} to T1 = .95 T_V or 1.05 T_V as per Quadrant
STAGE 2
Matching of Dynamic Stiffness in the Web Line
8
7
6
6
7

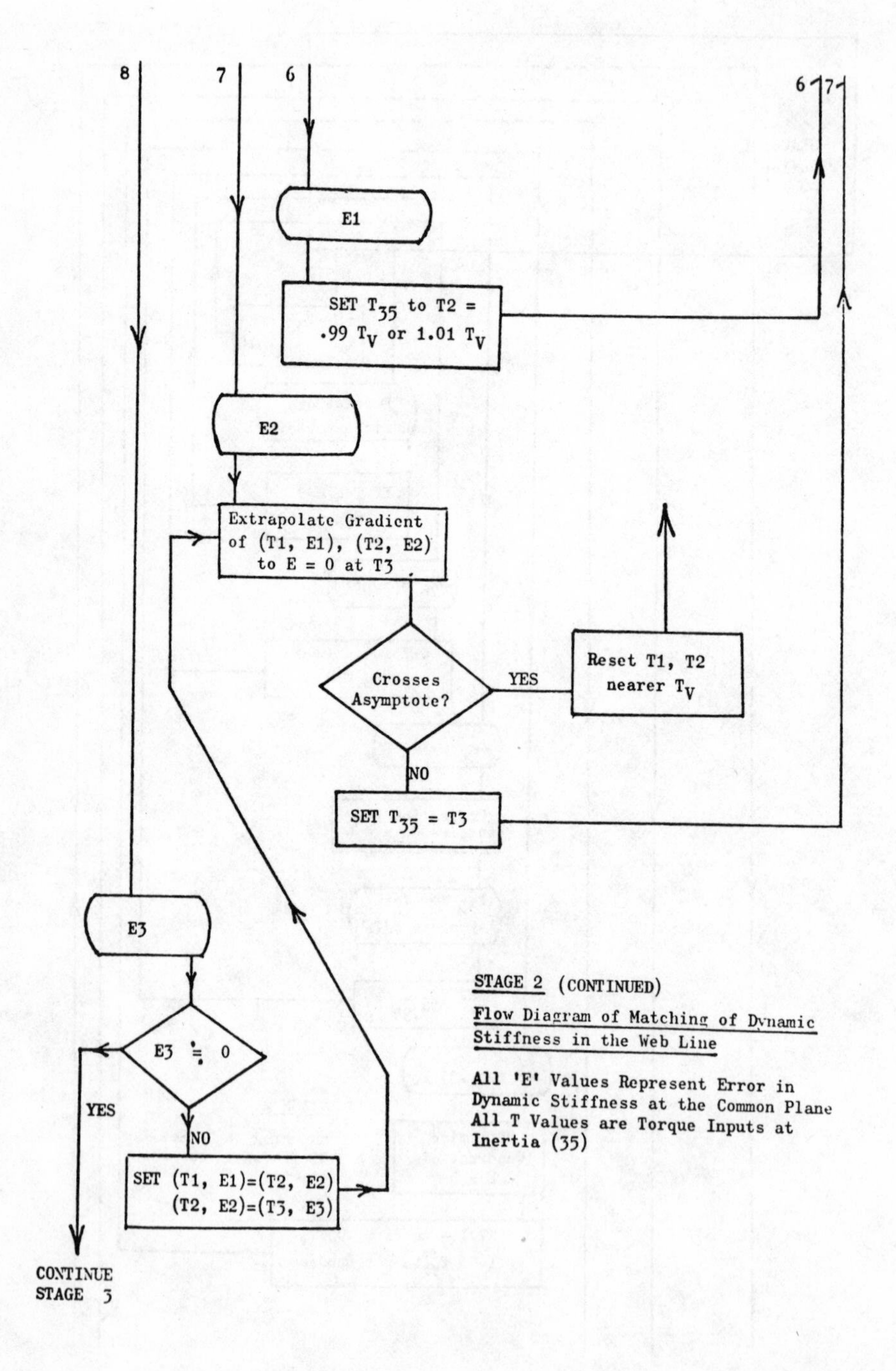

STAGE 2 (CONTINUED)

Flow Diagram of Matching of Dynamic Stiffness in the Web Line

All 'E' Values Represent Error in Dynamic Stiffness at the Common Plane
All T Values are Torque Inputs at Inertia (35)

APPENDIX II - NOTES ON DYNAMIC STIFFNESS

The concept of dynamic stiffness is an elegant adaptation of the Holzer tabular method of solution for eigenvalues of torsional vibration, applied to the matching of coupled systems, as, for example, in the matching of engines to shaft/propeller systems.

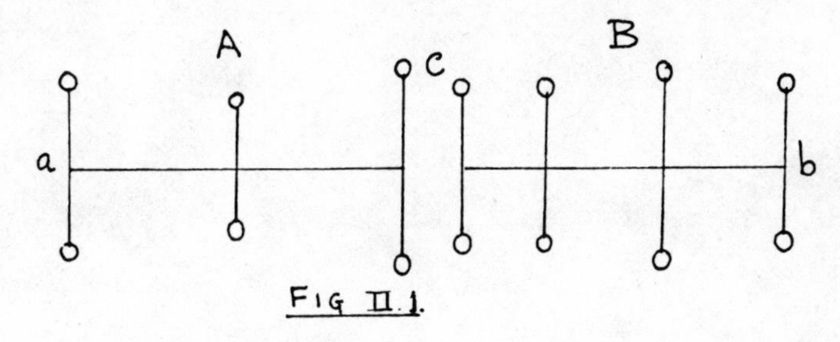

FIG II.1.

In Fig. II.1, it is required to find eigenvalues of the coupled system. Selecting a particular frequency, a Holzer calculation is carried out on system A, starting from end a, and a similar calculation is made for system B, starting from end b. Each system (except at the eigenvalues of the separate systems A and B) will produce a residual torque and displacement at C.

$$\text{dynamic stiffnesses} = \frac{\text{residual torque at C}}{\text{displacement at C}}$$

may now be calculated for each system. The advantage of using dynamic stiffness during matching is that this ratio is independent of the magnitude of the arbitrary inputs, since residual torques and displacements are proportional at a or b. If, then, the dynamic stiffnesses at C should be found to be "equal" from each side, this means that a simple proportioning of torques and displacements in <u>either</u> system A or system B will secure matching of torque and displacement at plane C. In such a circumstance, we have found a total system AB, vibrating at the stated frequency, with zero input torque at a and at b. This determines an eigenvalue frequency. To determine eigenvalues, then, it is only necessary to determine, by trial or iteration, those frequencies at which the dynamic stiffnesses at the junction are equal and opposite.